WOLF CONFLICTS

INTERSPECIES ENCOUNTERS

Series Editors: Rebecca Marsland, Alex Nading and Chrissie Wanner,
University of Edinburgh

The last decade has seen significant theoretical advances in critical animal studies, post-humanism, science and technology studies, perspectivism, and multispecies anthropology. This groundbreaking new series will solicit innovative works in the social sciences which have taken up the challenge to engage across species boundaries: humans, animals, insects, plants and microbes. On the strength of this work, the series seeks to expand methodological and theoretical approaches in the course of ethnographic engagements with other species. Questioning the distinction between human and non-human through innovative narrative and methodological strategies, books in the series will address a range of pressing social and environmental issues.

Volume 1. Wolf Conflicts: A Sociological Study.
Ketil Skogen, Olve Krange, Helene Figari

WOLF CONFLICTS
A SOCIOLOGICAL STUDY

Ketil Skogen, Olve Krange, and Helene Figari

berghahn
NEW YORK · OXFORD
www.berghahnbooks.com

First published in 2017 by
Berghahn Books
www.berghahnbooks.com

Library of Congress Cataloging-in-Publication Data

Names: Skogen, Ketil, author.
Title: Wolf conflicts : a sociological study / Ketil Skogen, Olve Krange, and Helene
Figari.
Description: New York : Berghahn Books, 2017. | Series: Interspecies encounters ;
volume 1 | Includes bibliographical references and index.
Identifiers: LCCN 2016054920 (print) | LCCN 2017007899 (ebook) | ISBN
9781785334207 (hardback : alk. paper) | ISBN 9781785334214 (eBook)
Subjects: LCSH: Wolves—Norway. | Wolf populations—Norway—Management—
Sociological aspects. | Wolves—Sweden.
Classification: LCC QL737.C22 .S599 2017 (print) | LCC QL737.C22 (ebook) |
DDC599.77309481—dc23
LC record available at https://lccn.loc.gov/2016054920

British Library Cataloguing in Publication Data

A catalogue record for this book is available from the British Library

ISBN 978-1-78533-420-7 (hardback)
ISBN 978-1-80073-178-3 (paperback)
ISBN 978-1-78533-421-4 (ebook)

CONTENTS

✳ ✳ ✳

PREFACE

✳ ✳ ✳

This book is the result of fifteen years of sociological research on the conflicts over wolves in Norway. As in many other countries, wolves and wolf management are contentious issues. The obvious connections between these controversies and other societal conflict lines have made them fertile ground for sociological research. There is potential for greater insight not only in challenges to wildlife management but also in fundamental aspects of modern societies. When wolves returned to places where they had been absent for decades, or even centuries, they became trapped in an already-existing web of social tensions. More than human-wildlife conflicts, what we see are social conflicts: they are conflicts *between* people *over* wolves.

Various parts of our research have been presented—during these fifteen years—in scientific articles, reports, popular pieces, public talks, lectures, and media coverage. Except for the articles, most of the dissemination has naturally taken place in Norway. But not all of it: Our studies have been presented at numerous conferences and workshops in North America and Europe. We have done comparative research with colleagues from France and India. Our findings are regularly discussed with colleagues from different parts of the world, most of them from academia but also practitioners from wildlife management and NGO representatives. Almost without exception, people nod in acknowledgment as they recognize the picture we paint of wolf conflicts as driven by social change in rural areas, embedded in class conflicts and struggles over knowledge, legitimacy, and power—even if the conflicts they know best may be about tigers, elephants, turtles, or, for that matter, logging and mining.

We are confident that this book has something to say to readers everywhere and that our examples, almost exclusively from Norway but with a little dash of France, are easily recognized by people who have experience with conflicts over land use, conservation, and wildlife management. The same goes for academics who have studied social and cultural tensions related to shifting class constellations, particularly in rural areas but even in urban settings.

The research we present is not brand new. Most chapters are based on texts published in scientific journals—rewritten and extended to become parts of a coherent whole—but some chapters contain material that has never been published in English. *Wolf Conflicts* is not an anthology. The most significant contribution is the overarching perspective that becomes much clearer in a book dealing with many aspects of our research, compared to what emerges from smaller pieces of text often dealing with one such aspect at a time.

We have written a book that we hope will work as a sociological essay but also provide new insights to people with a general interest in conflicts over wolves, wildlife management, and conservation. Finding this balance has not been easy, but we hope we have succeeded. Our research has been part of several projects funded from different sources. The major ones are the Research Council of Norway and the Norwegian Environment Agency. The Norwegian Non-fiction Writers and Translators Association and the Fritt Ord (Free Word) Foundation supported work related to the book itself. Our own institution, the Norwegian Institute for Nature Research (NINA), has also been supportive, financially and otherwise.

INTRODUCTION

✳ ✳ ✳

One evening in late October 1999, the people of Sørskogbygda, a small community in southeastern Norway, came out in force for a meeting at the community hall. The topic for debate was large carnivores, particularly wolves. At least two hundred people had crammed into the hall. On the stage sat a panel of three: Grete Fossum (member of Parliament for Labor and a local resident), Torstein Bilet from Our Carnivores (Foreningen Våre Rovdyr, an influential conservation NGO), and Olav Høiås, representing the regional environmental authorities. Torstein Bilet really was in the proverbial lion's den. Fossum's sympathies lay with those of most of the audience but not, it must be said, with those of her party. She was an adamant opponent of having large carnivores in the region. Høiås explained his role as that of a government official. He was impartial and did only what others decided he should do. He was an instrument of the state. The hall boiled like a cauldron.

Fossum proved a virtuoso populist. She got the audience on her side and won the debate. Many disagreed with Bilet, but that was expected. He and his organization want large, viable carnivore populations in Norway. When he said he might not oppose hunting if the populations were sufficiently robust, many were pleasantly surprised. Although the majority disagreed with Bilet, what he said was not especially provocative, because they knew where he stood. Most of the public's ire was reserved for the civil servant, Høiås. There were probably many reasons for this. First, no one believed in his assurances of impartiality. A government official with no powers? Inconceivable. Nor was the public ready to accept at face value what he called scientific evidence. For many in the audience, scientific results are merely opinions—more precisely, pro-wolf propaganda. Political statements served up as objective, neutral, and indisputable facts will obviously provoke people who disagree with them. The civil servant from the regional government agency was immediately branded as a friend of the wolf. His excuse for his self-proclaimed powerlessness—he was bound by government and parliamentary edicts along with international treaties—was blown out of the water by Fossum. Being a member of parliament gave her some credibility. The only thing to do, she told the public, was to carry on fighting—the battle against the wolf can be won. There are no two ways about it. Høiås represented the real enemy—the NGO Our Carnivores does not control policy on wildlife management: that is the authorities' job.

The positions that emerged during the debate were strongly anti-wolf, strongly pro-wolf, and a sort of middle-of-the-road, "responsible" position. And the enraged public primarily targeted the latter position, that of the authorities. We have seen distrust of authorities and of the science used by the government to support policy many times since that memorable evening in Sørskogbygda.

(Field notes from a public meeting in Sørskogbygda, October 1999)

This particular meeting happened at an early stage of our work on wolf conflicts as a subject for sociological study, and it taught us a few important lessons. First, the mere mention of carnivores can provoke a powerful response, as the highly vocal opinions expressed at the meeting showed. Large carnivores, wolves in particular, divide opinions and create conflicts. Second, as we saw, scientific evidence does not always come across with the authority the scientists themselves and the authorities would like it to have. The audience railed against the scientific evidence, dismissing any attempt to present it as neutral, objective, and independent. Science was seen as disguised political opinions, and scientists were actively taking sides—at least, that was how most of the audience that evening saw it. The third important lesson was that the government, the parliamentary majority, and nature management bodies are the real enemies of the anti-wolf camp. The government, joined by the carnivore biologists, was to blame for the return of the wolf to Norway. This gave us an early glimpse of an essential aspect of the dispute: hierarchical social structures and power. Anti-wolf campaigners in the hall resented and resisted the government's wolf policy that, in their view, powerful players in the higher echelons of Norwegian society, far away from Sørskogbygda, were imposing on them.

The community meeting is a good example of what we have seen throughout our work on wolf disputes. The struggle is about much more than a disagreement on the carnivore management regime and the practical consequences of having wolves in the vicinity. The public directed its anger at the government representative. Since that noisy meeting, we have talked with a lot of people who were clearly more annoyed with the government and the biologists than with the wolves. And here we already approach the book's thematic center of gravity. The wolf may well make a nuisance of itself and cause problems for farmers and hunters, but it has also had the misfortune of landing in the middle of historical social cleavages that run deep in Norwegian society. The battle over what counts as reliable evidence—that produced by the scientists and used by the government or lay, practical knowledge accumulated over generations—transcends policy areas. It is not restricted to the issue of carnivores or even to the question of wildlife conservation or land management. Essentially, it is about power relations and how people perceive the world from different positions in the social order.

Opponents of the wolf dominated the meeting in Sørskogbygda. For several reasons, the same could be said, to a degree, of this book. Opponents are highly visible in many of the areas inhabited by carnivores, and they frequently leave their mark on the local landscape of opinions. The economic and practical features of these conflicts, such as those involving killed sheep, are easy to see and understand. For our part, though, we concentrate on other sides of the dispute that come to the fore in places where the loss of livestock is not the core issue but where there is a lot of noise all the same. This is precisely the situation in the areas where we find wolves in Norway today. We shall attempt to explain why groups with strong roots in traditional land use, and often in a working-class culture, make up the nucleus of the wolf resistance. As a rule, they are not landowners or farmers, but we think they deserve an attention they seldom get, neither in the debate on large carnivores nor in other areas of political exchange. We would go so far as to say these groups in particular can help us understand how wolf conflicts—and important features of conflicts over large carnivores in general—are woven into relations of class and power, as well as processes of change in Norwegian society. We also aim to describe these processes of change, and their effect on the controversies over wolves, in such a way that they are easily recognizable in other parts of the world, especially in the Global North and on both sides of the Atlantic.

However, we have also observed that a lot of people see the wolf in a very different light. We have spoken with many who accept or even welcome the return of the large carnivore. Some farmers and hunters have a pragmatic view of wolves. Most important, though, is the growth in population segments with no strong ties to traditional land use practices, also in rural areas. These changes have an impact on people's relationship to the natural environment and their opinions on how it should be managed. It is in these groups we most commonly find a positive attitude toward the wolf. But, intriguingly, we also encounter a type of hostility toward wildlife management agencies that in many ways resembles what we have seen among wolf opponents, an observation that arouses our sociological curiosity. It goes without saying that some people are against wolves for obvious and, sociologically speaking, trivial reasons. The wolf is the cause of tangible problems for them. But this too can merge with other conflict dimensions in different ways. Our task in this book is to address all the different strands and present a more coherent picture.

The following point is vital, however: the book deals mainly with what happens in places where the wolf roams—how people there think and act, how the wolf for them finds a place in established ways of understanding reality, and how this is integrated in wider social contexts. The book is not a comprehensive assessment of Norwegian carnivore policy and management and says little about the institutional levels and large carnivores as a matter for government policy. We do not discuss the international treaties, such as the Bern Convention, under which Norway commits itself to conserving all species native to Norway, including the wolf. The book does not aim to describe—much less analyze—all aspects of conflicts related to large carnivores in Norway. For example, livestock loss is a pivotal issue in conflicts involving other carnivore species (brown bears, lynx, and wolverines) in parts of the country where rough grazing of sheep or reindeer husbandry is more important than in the wolf areas. Scientists—wildlife biologists—do not get to play a leading role, nor do conservation organizations. They are certainly included but primarily as part of the background against which the conflicts play out in wolf territory. Scientists and managers do get their say in some interviews, but most of the time they remain part of the context. Later in this introduction, we provide a brief description of large carnivore management in Norway, and we present some results from biological studies of the wolf in Scandinavia. We also explain why Norway currently maintains low population goals for all protected carnivores and why this does not significantly affect our analysis of the wolf conflicts. All of this is meant to fill in the background for the story we really want to tell, about the people who live in the wolf areas of southeastern Norway and how they look upon a rapidly changing world—changes symbolized for many of them by the wolf, for better or worse.

ON THE RETURN OF THE WOLF

The wolf returned to Norwegian forests in the 1980s. Although the last known individual from the original population was killed as late as the 1960s, Norway did not have a wolf population in any meaningful sense in the latter half of the twentieth century. When the wolf came back, it received a mixed welcome, to put it mildly. Almost immediately, sharp lines of conflict were drawn.

Large carnivores had divided Norwegians' opinions before the return of the wolf. Bears, wolverines, and lynx had been causing problems for sheep farmers and reindeer owners for decades. The controversy over large carnivores is a long-lasting cleavage. But while the conflicts have been with us for years, they escalated to new heights with the return of the wolf—not only in areas with wolves or other large carnivores but also in political circles at the national level and in the national media—for several reasons. One, of course, is what the wolf means to people. For some it means real problems, but for others it stands as a powerful symbol of wild, pristine nature. The carnivore question obviously informs the urban-rural relationship and is therefore drawn into major issues such as centralization, depopulation, and a general shift in the balance of power between rural and urban areas. These are burning issues throughout the world but not least in Norway, where rural policy has been very active in the postwar years.

Based on the public debate on carnivore management, one would think that almost all rural people are against the government's current policy on large carnivores: rural people appear to want smaller populations, or at least no increase. They are presented as the ones with firsthand experience of killed livestock, decreasing game populations, and the sense of personal fear. But what carnivore policy really affects, it is often said, is the sheer quality of rural life. At the same time, the idea of protecting carnivores is said to be typical of people in urban areas, whose idea of nature is often romantic: they want it to look like the wilderness on television and to be where they can enjoy outdoor recreation. From a certain rural perspective, these values appear to be on the offensive, as can be seen from the growing number of protected areas and in the management of carnivores, based on the central principle of protection of all species.

But many who identify with a modern and possibly urban culture see all this from the opposite perspective. Out in the country is where primitive peasants and hunters live: people too insensitive to appreciate biological diversity and wild nature, who only think of making a buck and getting as much as possible from the natural resources. As usual, the media fan the flames of conflict and provide ample space to both perspectives on the carnivore conflict as a purely urban-rural conflict. We would not be revealing any secrets if we called this presentation both an oversimplification and misleading. In this book, we shall explain why.

Norway's rural policy has been based on an understanding of agriculture as essential to the survival of rural communities, and this has been emphasized more here than in most other countries. Agricultural interests and organizations in Norway are powerful and exert a considerable influence on national policy and the wider political agenda. Given that livestock production in many areas is based at least partly on rough grazing, the industry can hardly be expected to extend a welcome to large carnivores. Sheep and cattle are vulnerable to attacks when they graze without supervision, and the same goes, of course, for the semidomesticated reindeer. The Sámi reindeer industry is naturally preoccupied with carnivore-related questions, which has helped turn the carnivore issue into a sizzling political topic that also touches on aboriginal rights. As we discuss in chapter 3, these powerful forces have also managed to define the carnivore problem as the property of the grazing industries and have succeeded in equating farming interests and wider rural interests (and the reindeer industry with Sámi interests). The equation is misleading, and it is one of the main themes of this book.

Many people outside agriculture worry that large carnivores kill game and that wolves also attack and kill hunting dogs, issues that are seen as threats to hunting, particularly in areas with wolves. Many people are anxious about being outdoors or even afraid that they risk meeting a wolf or bear, which can fuel negative perceptions of carnivores. But after our years of research, we are convinced these sentiments only explain parts of the opposition to the current carnivore management regime.

In this book we shall concentrate on the wolf, for several reasons. First, the wolf is an even more controversial species than the other three large carnivores in Norway (bears, wolverines, and lynx). Therefore, conflicts centered on the wolf are particularly good at shedding light on certain important patterns. In areas with a more or less permanent wolf presence today, there are not many sheep, and there is no reindeer husbandry at all. In fact, wolves in Norway have not killed many sheep (or other livestock, including reindeer). Nonetheless, conflicts are at least as intense as in places where farmers have actually lost large numbers of livestock to lynx, wolverines, and bears. Farmers make up a tiny part of the population in Norway today. Even in rural areas, most people work in nonagricultural jobs. The center of gravity of anti-wolf sentiment is found in groups outside the agricultural sector but who nonetheless identify with traditional land use and resource extraction. Furthermore, the primary enemies are no longer the predators themselves but rather people who favor larger carnivore populations in Norway. Carnivore conflicts have one important thing in common with other conflicts: they are conflicts between people.

Having wolves in Norwegian forests is simultaneously new and old. Humans and wolves have been in conflict from time immemorial, while conflicts between people over the wolf are relatively new. According to much recent evidence, the cleavage is becoming a permanent feature of what we might call the "rural state of affairs" in Norway. The reason can be found in social cleavages that penetrate far deeper than any dispute concerning carnivores. Since 1999, we have been interviewing individuals and groups about their attitudes toward large carnivores and their experience with them. We have interviewed several hundred people in the counties of Hedmark, Akershus, and Østfold—not only about carnivores but also about work and everyday life, future prospects, and the social structure of local communities. Everywhere we have been, we have heard more or less the same stories, and the conflict patterns have been the same. We are thus convinced that these controversies are not merely local and so we have a more universal story to tell. Studies of wolf conflicts offer a platform for saying something of wider significance about processes of change in rural Norwegian communities and modern societies in general.

As we have mentioned before but must stress again, wolves have real consequences for people. Only a tiny fraction of all lost sheep can be blamed on wolves, but they can ravage some herds if given the chance. Farmers receive economic compensation for lost animals, but they are also concerned because the animals suffer and attacks create practical problems in running the farm. The animals need more attention, breeding plans are disrupted, and so on. The media have published many images of mutilated sheep, with their legs and udders ripped out, some of them still alive. Owners say they suffer with their animals, and there is no reason to doubt them.

Hunting with dogs is a strong tradition in Norway. Dogs are trained to hunt birds, moose, and hare. In Scandinavia, the use of untethered, free-ranging, but highly trained dogs is also the norm for big game like moose. Now, hunters are increasingly wary about hunting with dogs in areas with wolves because the dogs are put in danger. Since wolves returned to Norway and Sweden a few decades ago, they have attacked several hundred hunting dogs and killed many of them. Such assaults are disastrous for the hunters, both because they love their dogs and because working together with the dogs is more important for many of them than the outcome of bagging the game. The time it takes to train a good hunting dog, as well as the fact that many dogs represent valuable breeding stock, does not make it easier.

People frequently tell us they are afraid of wolves and that this fear affects the quality of life in wolf areas. They may not let their children go to school alone, and old people are said to be afraid to go out and pick berries. These are tangible consequences. As long as sheep graze the land, game is to be had in the forests, and hunting dogs are on the loose, the consequence of having wolves in Norway will be the loss of livestock, reduced hunting success, and killed dogs. All of this represents a crucial and substantive—though not the only and possibly not the most important—reason why the conflicts remain as stable as they do. The number of wolf opponents and the temperature of the conflicts cannot be explained by the wolves' material impact alone. We therefore need to take a closer look at how people understand the arrival and presence of wolves in particular ways, and we do so against a backdrop of processes of societal change strongly felt in rural areas.

Before moving on, we need to stress again one of the main findings of our studies: local opinions are extremely diverse. In this book, we pay the most attention to opponents of the current management regime, who are extremely visible in many communities near carnivore habitats and likely to put their distinct stamp on the local "landscape of opinions." As mentioned initially, we have also interviewed people who are pleased the wolf is back. Importantly, however, they do not usually belong to the same social groups as the people who want the wolf removed, although there are exceptions. Quantitative studies (surveys) complement the picture. Arild Blekesaune and Katrina Rønningen (2010) found that Norwegians who live on a farm are more likely to dislike carnivores than others are. Also, independent of where one lives as an adult, a rural childhood tends to correlate with a skeptical attitude toward carnivores. But factors such as income, education, and access to cultural resources (often termed "cultural capital") also correlate with views on carnivores—the higher the score on such measures, the more likely it is that a person will accept carnivores, regardless of where that person lives (Blekesaune and Rønningen 2010; Skogen and Thrane 2008). The same pattern was found in Sweden (Krange et al. 2017). Tangible problems created by carnivores are important, but other factors have a powerful effect on opinion formation. By and large, surveys tell us that a considerable proportion of those living in wolf areas have positive opinions of the wolf and other large carnivores (see Tangeland et al. 2010).

In our qualitative studies, opinions vary within all social groups, but the studies have primarily revealed a tendency for people without cultural roots in traditional land use and the resource economy to express a positive attitude toward carnivores. Like their more skeptical neighbors, they are often deeply attached to nature where they live, but they have nothing against seeing it as a wilderness where humans play second fiddle and large carnivores naturally belong. Some of these people are hunters, but their outdoor activities often have nothing to do with harvesting. Like many of those with a traditional view on land use, they have often chosen to settle—or stay—in rural areas because of the natural environment. We have also seen an effect of social status, education, and cultural orientation: A "middle-class culture" appears to predispose positive attitudes toward carnivores, in rural as well as in urban areas.

Also important is to emphasize that many people, including in rural areas, have no interest in the carnivore issue at all. We have never had an opportunity to assess the degree of engagement statistically, but we would not be surprised to discover a silent and indifferent majority in many communities. Our clear impression, though, is that people who make little use of their natural surroundings are less likely to care about the carnivore question. And quite a few are in that category, also in rural Norway. In the following chapters, we will try to cast light on several aspects of the wolf conflicts as they play out in wolf country. One of the main themes in this book is that wolves have become entangled in conflicts deeply rooted in Norwegian society, indeedin all modern societies. We contend that the conflicts are about much more than the wolves and the actual problems they create. Livestock interests are there, but we claim they play a modest role. Our position stands in contrast, then, to what has become the prevailing discourse in politics, government administration, and the wider public debate.

WOLVES AND SOCIETY IN A HISTORICAL CONTEXT: THE ENEMY GAINS FRIENDS

The original Scandinavian wolf population was completely lost by the early 1970s, when the very last individuals had disappeared. In other words, it is entirely possible for humans to exterminate the wolf. Now that it has staged a comeback, the wolf lives here at the mercy of humans. What people decide to do is critical to the development of a carnivore population, and those who want wolves in Norway currently have the upper hand. This has not always been so. The fact that people perceive carnivores as a threat is nothing new. The pro-wolf mindset, however, that has been gaining ground is new. In this contradiction we find the potential for conflict today. It is therefore reasonable to say the aspects of the conflict that occur *between* people—the most important ones in our view—are also new and set modern conflicts *over* wolves apart from traditional conflicts *with* wolves.

An article published by Statistics Norway (SSB 2004) illustrates the point. The article, "From Bounties to Conservation and Irregular Killing" (Fra skudd-premier til fredning og irregulær avgang), shows that carnivores and carnivore management have been subjects of political dispute for a very long time. Norway adopted a hunting law in 1845, the Act Relating to the Extermination of Carnivores and Preservation of Other Game (Lov om Udryddelse af Rovdyr og Fredning af andet Vildt), to facilitate the hunting and eradication of carnivores without any value, that is, species whose behavior made them a threat to domestic animals and useful game. The idea was also that hunting wolves and bears would have other benefits. More than other forms, the hunting of carnivores required "bravery," "skill," and "perseverance" and therefore offered the best form of soldier training in peacetime. Certain types of carnivore, whose diet consisted mainly of snakes and rodents, were considered beneficial. The authorities did not introduce incentives to hunt badgers, for example. The same applied to numerous species of birds of prey. The fox, which could be a problem in the henhouse, had valuable fur, in addition to being a rodent predator. To introduce more incentives for the public to hunt foxes was unnecessary while bounties were used to encourage people to hunt and trap large carnivores. On the list of animals deserving to die—eagles, bears, lynx, and wolves—the wolf came first.

It must have been a successful campaign. According to Statistics Norway (SSB 2004), the wolf was almost certainly on the verge of extinction in southern Norway at the start of the twentieth century. As early as 1860, the amount paid out in bounties had fallen dramatically. In the same period, rural doctors were reporting a sudden fall in the number of children employed as shepherds. Larger livestock herds increased the area used for grazing in this period as well, probably because carnivores now represented a lesser threat.

More than 160 years have passed since the game preservation act, and as far as we know, the law, the bounties, and the goal to exterminate wolves and bears did not create even the smallest controversy. On the contrary, city people and rural folk all appear to have applauded the fight against carnivores. The men behind the law lived in Christiania (today's Oslo) and belonged to the absolute elite of Norwegian society. The bill was tabled by the liberal MP and university history don Ludvig Kristensen Daa and penned by Head of the University Museum Halvor Heyerdahl Rasch. Rural people did their bit by hunting and trapping. There is, however, evidence of a cleavage between center and periphery on another level, as the authorities became aware of widespread bounty fraud. Regarding the law's primary objectives, there was probably no dissent. We could say the law expressed a goal that center and periphery shared, which found support in all social strata. In those days, the carnivore conflict really was between humans and animals.

Unlike in the mid-1800s, this is only one aspect of the wider conflict today. Norway is not a country of farmers anymore. Farmers make up only a small minority of the population. They are also in a minority within the anti-carnivore camp. Opposition to carnivores is concentrated in groups at the margins of the primary industries but which nonetheless feel a strong sense of identification with the traditional use of the land and its resources. And the main enemy is no longer the carnivores themselves.

OUR SOCIOLOGICAL PERSPECTIVE

SOCIAL CONSTRUCTIVISM

This book addresses the controversies sparked by the wolf's return to the forests of southeastern Norway. Since conflict and sharp divisions of opinion often mark the situation, the subject here will be these conflicts and why they play out as they do. We have stated the obvious fact that wolf conflicts are conflicts between people. But as we have also seen, such was not always the case. Given this point of departure, we see two things. First, opinions about the wolf vary among individuals and groups. Second, opinions about the wolf have changed throughout history. From this emerges a delimited object of study—opinions about the wolf—created, or we may say constructed, by people. Our research concerns, then, the social construction of the wolf, and we consequently adopt the perspective of social constructivism.

To many people, a social constructivist approach may seem alien and controversial. Saying that nature is a social construction would strain the credulity of even the most benign biologist. An explanation is called for. The simple point is that all ideas about nature, including scientific ideas, derive from human thought processes, which never take place in a vacuum but rather in a particular social context. Collective ideas about reality can be of immense significance out there in the physical world. American sociologists William Thomas and Dorothy Thomas put it like this: "If men define situations as real, they are real in their consequences" (see Merton 1995). An example is often drawn from the 1929 economic crisis. Certain banks were rumored to be hovering on the brink of bankruptcy. People rushed to withdraw their savings, which in turn led to the actual collapse of some banks. Something similar happened in the United Kingdom during the 2009 financial downturn when the savings bank Northern Rock tumbled and fell. A third example is the impact of hoarding. If rumors say the stores are likely to run out of certain goods (because of a strike, for instance), they do run out. There may not be any immediate danger of shortages because of the strike itself, but goods become scarce because people act on what they believe. And the shortage of goods confirms, in many people's eyes, the truth of the rumor. Both the collapse of banks and the shortages in stores are examples of self-fulfilling prophecies, a special instance of a wider phenomenon: that people's interpretation of a situation leads to actions with extremely tangible repercussions. Therefore, to understand how particular interpretations of reality emerge is not only interesting but also important. Clearly, neither people's actions nor their thoughts lead carnivores to cause damage, but people's ideas about the wolf and its place in Norwegian nature guide their choices of action, the consequences of which may be large for wildlife management and for the wolf itself.

We can turn to another classic sociological contribution to explain what we mean, namely *The Social Construction of Reality* by Peter Berger and Thomas Luckmann (1967). Its perspective is wide, and a good deal of sociological literature begins by explaining the book's main points. Yet, the way of thinking advocated in the book represents a clear demarcation against the notion that reality can be observed and described "objectively." Since we are writing about a subject dominated by natural science, it might be useful to take a brief look at the book's main ideas. Readers well versed in sociological theory may find it odd to blow dust off this rather ancient contribution, but we find it useful for two reasons: it is considerably more pedagogic than most later contributions on the subject and—importantly—it draws a clear line between constructivism and idealism.

What we first need to explain concerns precisely the status that reality acquires when we adopt a social constructivist approach. This is crucial because it is a topic that tends to provoke reactions from people with a background in the natural sciences, such as biology. So let us say once and for all, constructivism does *not* imply idealism. Berger and Luckmann do not claim that ideas or thoughts "constitute" reality. On the contrary, they affirm the obvious and indisputable existence of a reality independent of the human mind. Nature, species, and ecological processes have real substance and exist completely independent of human consciousness. Employing a social constructivist approach is in no way in conflict with the science of biology, which studies and reveals nature as it exists in reality. But as far as meaning—opinions, understandings, and interpretations—is concerned, that is another matter. Meaning is created through social processes. Our ideas of the world are collective, shared by others—a banal prerequisite for communication. As there is necessarily a collective dimension to the production of meaning, we can say our understandings of all phenomena are socially constructed. The wolf, for example, has an existence completely independent of anything human, but the meaning the wolf has for us depends on our ideas and thoughts. The wolf is out there anyway, but our ideas of it are socially constructed.

That we must observe science from the same perspective can be discomfiting for us researchers. The scientific production of knowledge obviously has some important distinguishing features linked to the philosophy of science and methodology. But science has a lot in common with all production of meaning. Scientific studies are also conducted in the social world, and the knowledge produced emerges through social processes. The paradigms that at one point enjoy hegemony in science change over time. A simple example is the prevailing scientific view of the shape of the earth. It was once believed that the earth was flat, but today this view is marginalized, to put it mildly. Such changes are clearly influenced by the social, cultural, and economic context in which they occur. Therefore, scientific knowledge about a natural phenomenon, such as the wolf, is only one of several forms knowledge. There are many different opinions of the wolf in Norway today, and this is not simply a matter of being for or against. Knowledge about population numbers, dispersion, and behavior is produced in different ways in different social groups. Scientists' conclusions are challenged by other producers of knowledge, such as hunters and farmers on one hand and wolf enthusiasts on the other. Their type of knowledge can often be very different from scientific knowledge and has many adherents among lay people.

This brings us to another potentially controversial issue. By stressing the social origin of people's opinions, social constructivism leads to a form of relativism. We are asking how meaning originates in a societal context, for example, in different groups and institutions. The perspective per se is not of much use for ascertaining the veracity of different forms of knowledge, but it does not deny that some forms of knowledge do represent reality more truthfully than others. This might seem trivial but must be said, as there are many misconceptions on this point. A social constructivist perspective does not deny that at any given time a number tells us how many wolves are in Norway, that this number is correct, and that, consequently, all other numbers are wrong. The case may well be that scientists' population estimates are more likely to be correct than those of lay people, but that is irrelevant for the study of meaning production. Then we are interested in how different opinions of the size of the wolf population are formed and enjoy support in different social groups. All opinions need to be treated equally seriously. By not taking sides in such disputes, social constructivism may seem to relativize everything but only because its mission is not to uncover facts about wolf numbers, reproductive rates, and behavior. Put differently, a social constructivist perspective cannot be used to determine whether the earth is flat or round, but it does tell us that both ideas are created by people. This allows us to study the dispute between the two views of the earth's shape. We could try to say something about why one side was proved right without judging which of them *is* right. It is not certain that the side proved right is also actually right. These issues are usually strongly influenced by (shifting) power relations.

We may say that social constructivism according to Berger and Luckmann is a perspective on, not a theory of, mechanisms and causal relationships in society. The perspective points toward a specific type of research object, that is, those that concern meaning, such as knowledge, culture, laws, institutions, and power. Social constructivism, as we present it here, makes no assumptions about phenomena outside of human societies. Studies of the social construction of nature deal with the cultural significance assigned to nature, the institutions created to manage it, how laws are formulated, and how different segments of the public act in this social landscape of meaning and power. But the perspective per se says nothing about who gets power, which opinions and forms of knowledge get to dominate, or what counts as true. To do that we need other theories that say something about the links between people's interpretations of reality and what actually happens out there—for example, how different opinions about carnivores are linked to power structures in society and certain aspects of societal change. This is where our research contribution is located.

CULTURE

We have emphasized how the production of meaning—the social construction of reality—does not unfold in the minds of isolated individuals. Production of meaning is essentially social because language and the concepts we think with are social and because all communication is necessarily social. Frames of understanding, values, concepts, and symbols are shared by larger or smaller groups, they have a certain stability over time (despite always evolving) and a form of internal logic. This brings us to the concept of "culture." Culture, says the British cultural theorist Stuart Hall, is "the actual grounded terrain of practices, representations, languages and customs of any specific society. [It also means] the contradictory forms of common sense which have taken root in and helped to shape social meanings" (1996: 439).

Culture, then, is made up of collective social constructions—culture is basically a shared understanding of reality. Culture is not a separate sphere of social life, limited neither to literature and art nor to national costumes and culinary traditions. Culture is simply the dimension of meaning and interpretation in all social life; it is present virtually everywhere, in signs and symbols that all members of a group (maybe even a whole society) can understand. Language is one such system of symbols, as are traffic signs. Opinions of right and wrong, pretty and ugly, important and trivial—culture impregnates every aspect of our everyday lives. Paul Willis, another British cultural theorist, says, "It is one of the fundamental paradoxes of our social life that when we are at our most natural, our most everyday, we are also at our most cultural; that when we are in roles that look the most obvious and given, we are actually in roles that are constructed, learned, and far from inevitable" (1979: 185).

Culture is always shared, but the size of groups sharing particular cultural traits can vary considerably. Some culture elements may be shared by a whole nation, or even larger entities, while others are limited to smaller groups. When we talk about cultural differences, it is easy to think about differences between large categories, often nations or ethnic groups. Norway's culture is different from India's. However, it is obvious that India with its 1.2 billion people spread over a subcontinent is culturally heterogeneous. But so is Norway, despite its much smaller population and size. Since interpretations that create meaning constitute the essence of culture, culture obviously needs something to interpret and invest with meaning. What would that be unless it was precisely the world as it appears from people's own vantage point? Obviously, if the reality surrounding people looks different, as it will to people in different social positions, it will also result in different cultural patterns and sometimes in cultural conflicts: different modes of understanding or values are pitted against each other. Culture evolves as a consequence of changes in people's material conditions insofar as what people need to understand, explain, and relate to is also evolving. Clearly, culture can also influence people's actions, which in turn affects the material and social conditions of their lives; culture can lead to change but cannot be understood independently of such social conditions. In this book, the intersection of social class and place of residence (urban or rural) is particularly important when we examine the relationship between culture and socioeconomic context (and, of course, the consequences for the conflicts surrounding the wolf). Economic and social processes of change that are sweeping across rural areas, but that originate in globalization and what we might term general economic modernization, constitute a crucial backdrop to the cultural friction we can observe.

THE LARGE AND THE SMALL

Studies of culture and social constructions touch on one of the more central themes within the social sciences, namely the relationship between interpersonal relations on one hand and large societal structures on the other. As an everyday experience, the two levels are usually completely separated. We live our lives in interaction with family, friends, and colleagues. We also randomly interact with strangers, such as people on the street and assistants in shops. But this type of micro social interaction takes place within a larger, dynamic societal structure. This has to do with how the economy is organized, how work is regulated, national and international politics, institutions such as the educational system, and changes that occur at this highly aggregated level.

Interactions at the micro level, as well as the meaning and knowledge created there, are not independent of changes at the macro level. Thus, an important task for social science to understand individuals' actions and attitudes in the much bigger context, in relation to macro structures and change. The subject of wolves is no exception. Our studies have followed precisely such a program. We wanted to see opinions about the wolf in relation to the narrower as well as the wider contexts of the lives people live out there. When we look at the conflicts about the wolf against the background of modernization, class antagonisms, cultural conflicts, and power, this is precisely when we can hope to grasp a wider and deeper understanding of what is taking place.

The large and often global processes of change constitute the wider context of wolf opposition. Although we have studied local communities, our empirical data are influenced by general structural changes in society. Our contention is therefore that historical changes at highly aggregated levels affect the local and indeed the interpersonal. That is to say, wider global processes change social structures and social relations in rural Norway. The entire Western world has undergone economic modernization as the center of gravity has shifted from primary industries and manufacturing to service provision. In Norway, we have also seen a growth in public service provision and employment up until now. As a consequence, there has been a realignment of the class structure. The working class peaked in the 1960s, and in recent decades we have seen a rapid expansion of a highly educated middle class. Today, this middle class exerts a powerful influence in many areas of society, not least in the field of environmental policy and wildlife management. This field has expanded alongside the increasing proportion of the population with higher education. The new middle class filled the new jobs, and the mindset of its members gained a powerful influence in many areas of politics and government administration. Environmental policy and wildlife management officials—that is, the people whose job it is to deal with the management of wolves—are no exception. The Ministry of Climate and Environment, the Norwegian Environment Agency, and County Governors' environmental departments are bursting at the seams with university-educated employees. The same can be said of the environmental officers in the municipalities and about many others working in government at different levels.

When the wolves returned to the forests of eastern Norway, they came to areas where these changes were strongly felt. We can take Stor-Elvdal, a municipality where we collected much of our data, as an example. Local population figures peaked in the late 1950s. Timber, the municipality's most important natural resource, gave work to many people in the forests and at the sawmills. There were also jobs in agriculture. The number of active farms was higher back then, and there were jobs in the local dairy. The national rail service was an important workplace, and the railway station was manned. Stor-Elvdal was a class society of the "classic" type, with a large working class that sold its labor in the resource-based industries. On the other side of the class divide, a few big forest owners dominated. Much of this has changed in the wake of general modernization processes. The property structure is more or less the same, but forestry technology has advanced tremendously, and the industry employs far fewer people. The dairy is gone, and the railway station is no longer manned. The population has nearly halved. The local council administration is the biggest employer by far, being responsible for schools and primary health care. Many public employees have manual jobs or jobs that require limited education, but the council also employs an increasing number of people with college degrees. Many in the latter group are newcomers to Stor-Elvdal. Despite the growth in the public sector (until recently), the council is fighting with its back against the wall. As the population declines, cuts must be made to vital social services such as education and health. Stores and gas stations are already gone. When the wolf turned up, it came to symbolize a development that comprises these unpleasant things, as well as changes in environmental policies that affect traditional land use, seen by many as part and parcel of the same trend. This certainly exacerbated a conflict that otherwise came to encompass tangible things such as the loss of sheep, dogs, and game.

Generally, what we have written above can be read as a model of how we work sociologically. For example, we wanted to understand people's opposition to wolves and quickly discovered it has to do with much more than the loss of livestock and game. In this sense, the study of the carnivore turned into a wider study of Norwegian society, a kind of focal point for a discussion of general social dynamics. That is why this book is about not only wolf conflicts but also important aspects of contemporary society. The two levels cannot be understood independently of each other.

HOW THE BOOK IS ORGANIZED

First, it is necessary to provide a basic understanding of the complexity of the conflicts over wolves and, indeed, the complexity of factors that lead to an apparent united front against current wolf management in some small rural communities. To this end, we present the somewhat fragile but sociologically interesting anti-wolf alliance in Stor-Elvdal (in chapter 3). Thereafter, we will quickly zoom in on some of the social mechanisms we see as driving the conflicts—beyond directly affected economic interests, which are clearly a strong force behind some groups' stance on the wolf issue but just as clearly absent in the case of others. We will show there are dynamics at play that—for the social groups situated in the core of the conflict—only indirectly, if at all, involve livelihoods at risk or other threatened material interests. Unless such dynamics are properly understood, the totality of the conflicts over wolf conservation—indeed, over conservation in general—cannot be fully grasped.

Social groups with different "stakes" in the issue may follow different paths to a stance on the specific issue of wolf management that unites them in a form of alliance and shared discourse that bridges other, sometimes deeper, tensions. Therefore, we cannot disregard these other bases for engagement. Economic loss and practical challenges are certainly the most visible driving forces, not least in the media and in public debate, but our task is to explain that they are not major factors in the popular engagement with what we may term wolf politics.

In this book, and in our research over the years, we have been preoccupied with social groups that in many ways bear the brunt of economic and social change in rural Norway. In our study areas, people with a working-class background and deep cultural roots in resource extraction and harvesting are the dynamo of the resistance against wolf protection. For the most part, they do not own land or sheep, and their anti-wolf trajectory is not identical to that of landowners and sheep breeders. As we shall see in chapter 3, the latter groups do not enter the "alliance" by the same route either. In chapter 5 (on social representations of the wolf), it gets even more complicated, as we introduce the new and positive ways to conceive of the wolf that are also present in wolf areas—but predominantly with a basis in still other social groups.

The larger context of economic and cultural change is present in various ways throughout the book. In the concluding chapter 10, we aim to draw the strings together and elaborate on the relationship between rurality, social class, and power. But first we turn to a brief account of the status of the Norwegian wolf population, the Norwegian management system, and the historical background for Norwegian large carnivore policy.

CHAPTER ONE

THE WOLF IN NORWAY

❅ ❅ ❅

What we say about wolves in this book is based on analyses of conversations with people and the opinions they expressed. They have opinions—often strong—about the wolf in Norway, but what concerns them most is not always the wolf itself. Even when talking about wolves, they often drift off to aspects of the wolf issue that seem most important to them, which regularly concern government, politics, management, and public spending. Or they talk about the dominating discipline in large carnivore research: biology. Clearly, many are also concerned about the animal itself. Its distribution, behavior, population size, and population changes are issues where opinions differ. Scientific knowledge about wolves will be a key topic in this book, but primarily as laypeople understand and relate to it. However, as background to a book dealing with the wolf's reception in Norway, we must give biologists a chance to describe the animal's population status. There is extensive biological research on wolves, and hardly a wolf population in the world has been mapped as thoroughly as the Swedish-Norwegian one. This is obviously because the conflicts surrounding it have created an enormous need for knowledge in government agencies, which in turn has released research funding.

After explaining the population status of wolves in Norway, we review the structure and organization of the Norwegian carnivore management regime, which is necessary because our informants often refer to management agencies and other actors of the Norwegian system. We will keep this presentation simple, as this is not a book about Norwegian wildlife management. However, most readers will surely recognize the types of actors involved. Our experience is that people familiar with wildlife management in their own countries immediately recognize the types of institutions, categories of professionals, and so on that fill approximately the same roles as those we describe here.

POPULATION MONITORING

Norwegian carnivores do not roam free and unobserved in the forests like genuinely wild animals—least of all the wolf, which is under intense surveillance. A large team of scientists monitors the Norwegian and Swedish wolves as closely as possible all year round but especially in winter when tracks in the snow are easier to follow. Only a few high-profile left-wing activists of the seventies may have been subject to anything like it in postwar Norway, and maybe a handful of Islamists today.

The border between Norway and Sweden cuts right through the wolf's habitat. In 2000, a joint Norwegian-Swedish scientific venture named Skandulv ("ScandWolf") was set up. The main purpose of this bilateral research project is to provide scientific knowledge for an optimal management of wolves in Norway and Sweden. "Optimal management" is no small ambition, and a quick look at Skandulv's website confirms how wide-ranging their work has been. At least once every year it issues a report detailing the official estimates of the Scandinavian wolf population, an important and resource-consuming part of the project. In addition, research findings on, for example, population dynamics, predation, inbreeding depression, mortality, illegal hunting, and migration are reported and to a large extent published internationally.

The dual objective of Norwegian carnivore management—viable populations with as little conflict as possible—serves as a background to the Norwegian effort. It is commonly held that to achieve these objectives, a knowledge-based management is required. Therefore, research is perceived as a management tool. The goal is the long-term survival of the wolf population (although current policy is to keep numbers low), and large resources have been funneled into knowledge production. The idea is not a strange one: if you want to manage a species so that it thrives and reproduces, you need to know as much as possible about it, especially if you want a small population that is still viable in the long run.

But whether research has contributed to the second objective—as little conflict as possible—is debatable, to say the least. Indeed, as we shall see, science has been part of the conflict in two ways. First, findings presented by the scientists are disputed. Many people in the anti-carnivore camp raise doubts about them; the population estimates, for example, are seen as too low. Even the activity itself is criticized, as scientific methods involve a significant level of interference with nature and wild animals. Motorized vehicles transport scientists and equipment into the forests, and to mount a GPS collar on a wolf, it must be immobilized, usually by shooting darts from a helicopter. The number of wolves wandering around with a GPS collar is considerable, and some animals have undergone the immobilization procedure several times. For some people, these operations amount to cruelty to animals. Others have problems with what they see as a form of domestication of wild animals. We will discuss these issues later. For the moment, we will just present some of the results of the monitoring program.

To monitor wolf populations is no easy matter, and scientists use several methods. Modern technology like DNA analysis, electronic tracking (GPS collars), and camera traps has become important, but snow tracking is still the cornerstone of population monitoring. Many institutions are involved in both Norway and Sweden, and staff from universities and research institutes, as well as management agencies, are involved. NGOs and individual volunteers also participate (and we will return to this particular aspect in chapter 6). In the late 1960s and early 1970s, the south Scandinavian wolf population became extinct (Vila et al. 2003). Some sightings of wolves occurred in the following decade, however, and no year was completely devoid of track observations. By the early 1990s, the population had begun to grow at an annual rate of about 25 percent due to immigration. Using comparative DNA analysis, scientists were able to identify the new animals as of Finnish-Russian origin. Until relatively recently, the entire population stemmed from three individuals, but newcomers have shown up in the past few years, and these too are from Russia.

In 2015, Sweden and Norway shared a trans-boundary population of approximately 450 animals. About sixty-five of these lived only in Norway, and about twenty-five more had their territories on both sides of the border. There were seven reproductions in the Norwegian packs in 2015, and three more in the cross-boundary packs—a record high in modern times (Wabakken et al. 2016). In a comprehensive report on mortality in the south Scandinavian wolf population, illegal hunting was indicated as the main cause of death. From 1999 to 2006, about half of all deaths were probably caused by illegal hunting, which is seen as the greatest short-term threat to the Scandinavian wolf population (Liberg et al. 2011). Limited license hunting is permitted in Norway outside the management zone for wolves (see next section). After having no legal hunting at all, Sweden has recently allowed for it, and on a relatively large scale: forty-four wolves were killed by Swedish hunters in 2015.

Wolf management is meant to build on an already-extensive body of knowledge. Considering the great effort to produce knowledge, the wolf would indeed appear to be in safe hands. The Scandinavian wolf population has grown steadily since the 1980s. Therefore, it could be seen as a paradox that the wolf is still on the Norwegian Red List and labeled "critically endangered." However, high mortality due to legal and illegal killing and partly low reproductive capacity explains why the population is still at a level that experts consider too low. Ironically, the low population goal set by the Norwegian parliament effectively guarantees that the wolf will never leave its status as critically endangered. Although the Swedish population goal is higher than the Norwegian one, the wolf is considered

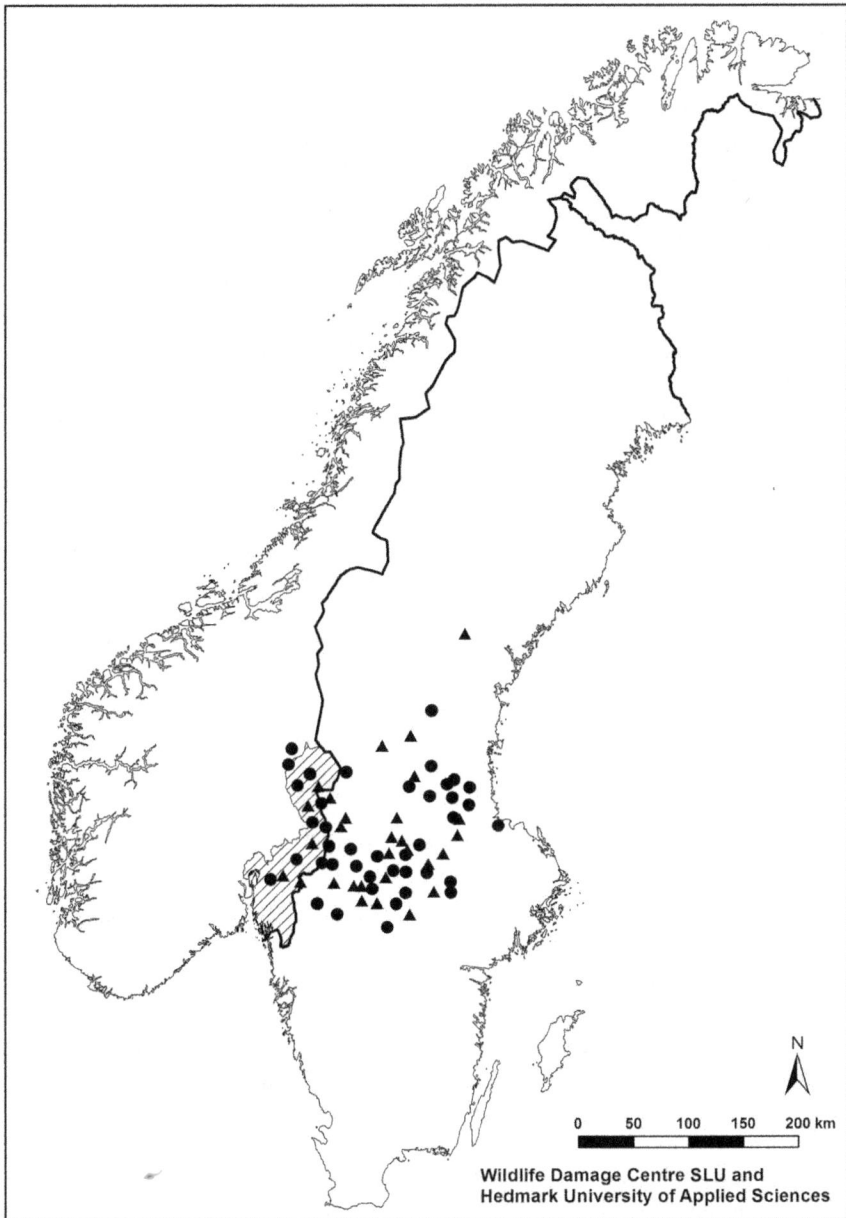

Figure 1.1. Wolf packs (circles) and territorial pairs (triangles) in Norway and Sweden during winter 2015–2016. The Norwegian management zone for wolves is indicated (hatched). Source: Wildlife Damage Center (SLU) and Hedmark University for Applied Sciences. https://brage.bibsys.no/xmlui/bitstream/handle/11250/2390916/3/Rapporten.pdf

"vulnerable," according to the International Union for Conservation of Nature (IUCN) criteria (Artdatabanken 2015; Liberg et al. 2005).

The Swedish-Norwegian wolf population is relatively isolated, and DNA analyses show that the population had just three founders and is therefore highly inbred. If an individual's inbreeding coefficient is 0.25, its parents are full siblings. As early as 2000, several individuals in the Scandinavian wolf population were already showing inbreeding coefficients of about 0.30 (Liberg et al. 2005). All of the individual wolves had more DNA in common than normal siblings, and scientists were concerned about the possible consequences. Inbreeding depression can lead to various diseases and low reproductive rates. Some indications of these problems have been recorded, for example, small litters and some birth defects (Räikkönen et al. 2006).

In 2008, however, the south Scandinavian wolf population received fresh genetic material for the first time in fifteen years. Two male wolves from Finnish-Russian areas fathered their own litters in 2008, and their genes have subsequently spread into a number of packs, improving the situation. Since then a few more wolves have arrived, and while this has done much to strengthen the gene pool, experts say it is still not enough. Swedish authorities at one point considered introducing individuals from the original populations in Finland and Russia, or taking cubs from zoos and giving them to wild wolf mothers. These ideas have been met, not surprisingly, with widespread resistance and are now shelved. Authorities have dismissed any thoughts of doing the same in Norway.

LARGE CARNIVORE POLICY IN NORWAY

Livestock production, though limited, does exist in the wolf area. Compensation programs are in place as in the rest of Norway, so farmers suffer no great economic loss even if livestock is killed. Extensive and costly preventive measures funded by the government—such as economic support to move sheep, erect electric fences, and so on—limit further damage to livestock. However, these measures are directed not specifically toward wolves but rather at large carnivores in general.

Extensive sheep husbandry can be found in other parts of Norway. Here, livestock loss does play a crucial part in the conflicts over large carnivores, and the same applies to the reindeer herding areas in the north. However, these areas do not currently have wolves: the culprits are lynx, brown bears, and wolverines. These depredation problems are also relatively new and a result of species protection implemented since the 1960s, but they started long before the wolves reappeared. Thus, conflict with livestock was already established as a core issue in Norwegian large carnivore management.

Norwegian regional policy strengthens the centrality of the farming per-spective in relation to problems with large carnivores. Maintaining rural set-tlement all over the country has been a stable political goal, and stimulating the agricultural sector and other resource-based industries (forestry, fisheries) has been the cornerstone of this policy (Almås 1989). A high level of subsi-dies—paralleled only by Switzerland, Iceland, Japan, and South Korea (OECD 2012)—has maintained active, technologically advanced agriculture spread across much of the country. However, most farms are small by European stan-dards, and in many areas, due to climatic factors, livestock and grass production are the only viable farming activities. Shifting governments have thus stimu-lated livestock production, not least sheep breeding. During the long period when large carnivores were effectively absent (the twentieth century up until circa 1975), herders developed husbandry methods entailing free-ranging sheep with limited supervision. Breeding deliberately weakened herding instincts so as to make the sheep disperse across mountains and forests to better utilize the grazing resource and prevent the spread of disease and parasites. Consequently, shepherding and the use of guard dogs are now difficult, unless older sheep breeds are used, which have a much lower meat yield, so herders are reluctant to go that route. When large carnivores returned, effects were extremely serious in some areas, and conflicts have flourished ever since.

Due to the active role of the state, a large agriculture bureaucracy exists and has its counterpart in large and strong farming organizations. The inter-action between the state and the farmers is well regulated, and numerous offi-cial communication channels exist. For example, farmers' organizations and the government annually hold extensive negotiations over subsidies and other issues affecting the agricultural sector. Consequently, there are well-established systems for transferring economic support from the state to the agricultural sector, and mitigation efforts are almost exclusively designed to help sheepherders—even in the wolf areas, where rangeland grazing is very limited.

A backdrop for our account in this book is what we may term rural decline: economic downturn, depopulation, service cutbacks, and so on. In Norway, the impact of this development is dampened by a regional policy that supports rural settlement and rural industries. While this still sets Norway apart from many other countries, government engagement in rural affairs has scaled down in recent decades, and globalization is definitely catching up with rural Nor-way. This development has a time lag compared to, for example, neighboring Sweden, but, seen from a rural perspective, we are now catching up, so there is widespread political frustration in rural areas. Yet, the political influence of rural interests, especially tied to the farming and forestry sectors, remains comparatively strong, which leads to a management regime for large carnivores that many would consider restrictive, with relatively low population goals for all species.

THE NORWEGIAN LARGE CARNIVORE
MANAGEMENT SYSTEM

Norway is presently divided into eight management regions for large carnivores. Some authority over management has been transferred to regional boards that are politically appointed (by the Ministry of Climate and Environment but based on nominations from the county[1] assemblies) in an effort to bring decision-making closer to the people affected and to introduce an element of "local democracy" to carnivore management. The ultimate goal is conflict reduction through increased legitimacy, a system implemented in 2005. Research indicates that many local people do not see the boards as either sufficiently local or under any meaningful democratic control. Furthermore, the boards are hampered by very limited powers within a system controlled from the national level, which renders them relatively irrelevant in the eyes of interest groups on both sides of the controversy (see chapter 8). These regional bodies (with technical and scientific support from the County Governors' environmental departments) are now responsible for managing brown bears, lynx, and wolverines, within a nationally established framework entailing, for example, exact population goals for each region and species set by the central government.

A special management zone for wolves has been established on top of this system, due to particular management challenges, including the virtual impossibility of combining territorial wolf packs and free-ranging sheep. This zone partially covers two management regions, involves four counties, and is managed by the two regional boards in collaboration. As one can see, this is a complicated system, not likely to alleviate people's sense of alienation toward management institutions (see chapter 8). Furthermore, the boards cannot go outside their specific mandate; if they do, the national Environment Agency will take over. Their decisions on hunting quotas, which they may set as long as regional population goals have been met, can be appealed and potentially overturned by the Ministry of Climate and Environment, which has happened on several occasions. So even though the boards are often regarded as predominantly "anti-carnivore" and "pro-farming" (no stakeholder representation, only political appointments), their establishment has not changed the management regime on the ground in any major way, as many local people have observed, which causes frustration for those who would like to see fewer large carnivores around (see chapter 8).

The general principle of the wolf management zone (see figure 1.1) is that wolves are allowed to establish territories and breed inside it, and the threshold for culling problem animals will be high. Outside the zone, however, this threshold is lower, and repeated attacks on livestock will trigger culling. As mentioned, there is also some very limited license hunting outside the zone;

Region 8 - Troms/Finnmark

Region 7 - Nordland

Region 6 - Midt-Norge

Region 5 - Hedmark

Region 1 -
Vest-Norge

Region 3 -
Oppland

Region 4 - Oslo/Akershus/Østfold

Region 2 - Sør-Norge

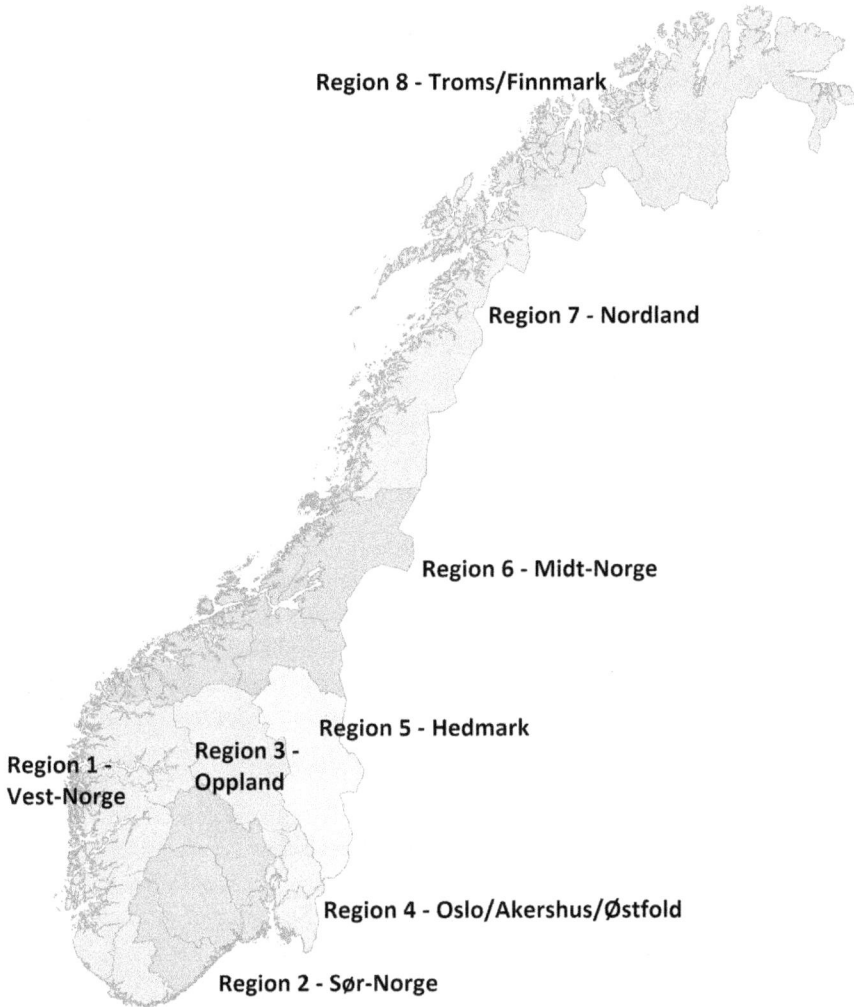

Figure 1.2. Management zones for large carnivores in Norway.

the quota was four animals in 2015. The management zone for wolves has been delineated specifically to minimize livestock losses, that is, by drawing up a zone covering an area with little livestock grazing. However, this has only been possible because the wolves are already concentrated in the border areas, close to Sweden. When the new immigrants first crossed the border, they arrived in a part of the country that is partly forest areas, where farming has never been economically important, and partly agricultural areas with plant production and limited livestock husbandry.

WOLVES: A SOCIAL PHENOMENON

Following this presentation, we can establish that the wild animal that is the wolf does not live a completely wild and free life. It lives here at the mercy of humans. We have divided the country into zones and erected fences; we hunt and kill "problem animals." We decide where and how the wolves can live. We count, measure, and weigh them and keep track of who mates with whom, how many cubs are born, and how closely related they are. We know where they live and where they roam. Many of them wear collars, allowing us to follow their every move and know where they are at any time via GPS technology. Increasingly, their social life is captured by camera traps and posted on Facebook—even as part of the communication efforts of scientists and managers.

Politics, management strategies, and individuals' actions decide how many wolves get to live and where they can stay. What humans do determines the population's genetic health. Carnivores have become a significant topic of politics, policy, and management, with dedicated items in the national budget, an elaborate and costly management system, and not least ardent political actors on either side of a conflict-ridden opinion landscape. The politically determined population goals have been reached in both Norway and Sweden. The wolf population, in other words, is as big as politicians have decided it shall be, but inbreeding depression is still considered a problem. Within the science and management community, the dominant "social construction" of the wolf in Scandinavia is that the population is small, isolated, and vulnerable. By using the term social construction, are we implying that it is *not* small, isolated, and vulnerable? No, but we emphasize that the wolf is a cultural, social, and political phenomenon. We did not like it when it was here before, so we exterminated it. Today, influential people and powerful institutions want it back, which is a crucial condition for its survival. The wolf is here at the mercy of humans—which returns us to the topic of this book: the sociological subject *wolf.*

NOTE

1. A Norwegian county (*fylke*) is a spatial and political unit at the intermediate level, between the municipalities (*kommuner*) and the national government. It has an elected assembly and is responsible for regional planning, transport, upper secondary education, culture, outdoor recreation, and harvestable game and freshwater fish. However, large carnivore management is the responsibility of the County Governor's office (*fylkesmannen*), which represents the national government at the county level.

CHAPTER TWO

AREAS OF STUDY
AND METHODS

❄ ❄ ❄

We have studied people's opinions about the wolf and the conflicts following in the wolf's tracks. We conducted fieldwork in areas of southern Norway where wolves have been present in the past two decades or are present today. These areas extend along the border to Sweden in southeastern Norway, from the municipality[1] of Stor-Elvdal in the north to Halden in the south, with Trysil, Aurskog-Høland, and Våler in between, and from the thinly populated forest communities in the county of Hedmark to the more densely populated agricultural areas at the rural-urban fringe in Østfold county. Below, we give a brief account of the population, economic structure, and carnivore status in the five municipalities making up our area of study. Following that, we say a little about the research methods we used.

Stor-Elvdal, a typical forest and logging municipality, lies in the middle of the Østerdalen valley. Just over 2,500 inhabitants share an area totaling 2,167 square kilometers, although roughly two-thirds live in Koppang, the municipal center. Forestry and related industries (sawmills, for instance) have always been the mainstays of the local economy. Livestock production dominates farming in Stor-Elvdal, and there are some rough-grazing sheep, especially on the western side of the municipality (toward the Rondane mountain range). The population fell from 5,470 in 1950 to the current level. The pattern of out-migration is one common to many other rural municipalities, and young people, especially women, are those who tend to leave. The primary industries have shed jobs in step with rationalization. And while more jobs have appeared in the public sector, they are not enough to balance the losses. In contrast to several other municipalities in the region, Stor-Elvdal does not have a significant tourism industry.

A small group of landowners is not only economically powerful, but controls ordinary people's access to their main recreation resources, hunting and fishing, since both are landowners' rights in Norway. The twelve biggest estates own about 55 percent of the productive forest. These property relations have created, and continue to create, clear divisions between people in different social positions. The state and the municipality itself also own large tracts of land, however, ensuring that people are not entirely dependent on the private landowners for an opportunity to hunt. The natural environment is probably still the municipality's greatest asset. Although the economic significance of Stor-Elvdal's natural resources has decreased, large forests and mountain ranges provide varied opportunities for outdoor activities. Hunting plays an important role, as it does in many other rural communities.

During the long period of the twentieth century when large carnivores were virtually absent from Norway, ungulate populations grew dramatically (not only because there were few carnivores but also because of successful management and changing forestry practices). This increase, together with the postwar growth in prosperity, turned hunting into the mass phenomenon we know today. As the recreational importance of hunting grew, so too did its economic impact. Leisure moose hunting, which is so important today, is a relatively young activity, but in many rural communities it is seen as an ancient tradition. Participation in the hunt is often considered an expression of bonding with the local culture (Brottveit and Aagedal 1999), including in Stor-Elvdal.

In addition to game such as moose, reindeer, roe deer, hare, and various birds, Stor-Elvdal is one of a very small number of municipalities with all four Norwegian large carnivores: wolves, brown bears, wolverines, and lynx. Wolves were not permanently present in Stor-Elvdal in 2015, and the municipality is not included in the current wolf management zone (chapter 1). Nevertheless, in 2000, Stor-Elvdal had two territorial wolf packs. The Norwegian Nature Inspectorate eliminated one of these, the Atndal pack, in 2003 because it had ventured beyond what was then the designated wolf area and started attacking livestock. The eradication of the Atndal pack was highly controversial, and friends of the wolf gathered from far and wide to stage protests and obstructions. These actions interfered with the inspectorate's work, forcing it to hunt from a helicopter, which created even more hostility. Although our study of wolf conflicts started in Trysil (in 1999), Stor-Elvdal was the first community where we got to know people over a longer period. We started working there in 2001 and followed developments keenly for several years. Stor-Elvdal is therefore a key municipality in this book, as will soon be discovered.

The municipality of Trysil lies as far east as possible in the county of Hedmark. To the east and south is the border with Sweden. The geographical and administrative center is Innbygda. The municipality covers a total area of 3,011 square kilometers, making it one of the largest municipalities in southeastern Norway. About 60 percent of the land is productive forest, and 9 percent is cultivated land. The rest is mountain ranges, low-productivity forest, and water.

Today, 6,800 people live in Trysil. Like most rural municipalities, the population has followed a downward trend since the war, falling by about 1,600 between 1951 and 2007. The population grew slightly in 2008 and 2009 but then resumed its previous pattern of steady decline. About half the population lives in semi-urban settlements or small hamlets. A large ski resort has been developed in the Trysil mountains, and tourism has become the leading industry. But with 1,600 square kilometers of forest, Trysil is also Norway's leading forestry municipality in terms of logged volume. Statskog (the Norwegian state-owned land and forest enterprise), the municipality itself, and various corporations together own 46 percent of the woodland, in contrast to Stor-Elvdal, where the large forest properties are family owned.

Livestock production is the main form of farming in Trysil, though local authorities say on their website that the conflict between carnivores and grazing animals makes it difficult to exploit the municipality's natural resources. The number of sheep farms has fallen dramatically. Several farmers have suffered drastic depredation but mainly to bears and lynx. Most of Trysil had no permanent wolf presence until recently, although at the time of our field work, residents tended to be very aware that they were "surrounded" by wolves in neighboring municipalities and in Sweden. A few packs had territories that extended into remote corners of Trysil, and stray wolves had been recorded for quite some time. The Slettås pack settled in the northwest in 2010. Trysil today, then, has a stationary wolf pack within its boundaries. Many species of game thrive in Trysil, and hunting is an important pursuit. Thirty percent of the male population paid the national hunting fee in 2009. With 1,143 felled moose in the autumn of 2009, Trysil is the biggest moose-hunting municipality in Norway.

Aurskog-Høland lies farthest east in the county of Akershus and shares borders with Sweden and the counties of Hedmark and Østfold. Aurskog-Høland is the biggest municipality in Akershus, with an area of 967 square kilometers. The major part of the population lives in and around the municipal center, Bjørkelangen. Our informants stem mainly from the more thinly populated areas of the municipality and do not represent as wide a section of the population as our informants in Stor-Elvdal and Trysil. They were originally selected to take part in a study of hunting, but most of the interviews featured issues with carnivores, including wolves.

Of the area covered by the municipality, 12 percent is cultivated land and 70 percent productive forest. Both figures are well above the national average. The population has been growing steadily since the turn of the century and today numbers more than 14,000. There is some manufacturing in Aurskog-Høland, but nearly one in four of the working population commutes outside the municipality, to Oslo or the nearby town Lillestrøm. Aurskog-Høland is simultaneously both a rural and suburban municipality. Grain production is the most important form of agriculture, and there is very little sheep farming.

Aurskog-Høland is also a major moose municipality. In addition to moose (and roe deer), people hunt a varied menu of small game. Aurskog-Høland has no permanent wolf presence today (although strays do appear regularly) but did a few years back—and conflicts were very pronounced at the time. Lynx are found in the municipality, but there is no permanent population of bears.

Between 1998 and 2003, the municipality of Våler (not far from the larger town of Moss in Østfold county) hosted the Moss pack. We conducted a smaller study in Våler in 2001, focusing on relations among regional wildlife management authorities, wildlife biologists, and local stakeholders (there was no cross-section of the population in this study). A comparison of the Østfold municipalities with those in Østerdalen highlights the different settlement patterns. Population density in the former is far higher, and municipalities such as Våler have much more arable land than Stor-Elvdal and Trysil, for instance (and most of Aurskog-Høland). Roads, fields, and buildings surround the small patches of forest. There is no wilderness at all, and wolves that settle here are never far from people. A half hour's drive is all it takes to get from Våler's town hall to the center of Moss. People in inner Østfold are not used to thinking about their local environment as a habitat for large carnivores, even though outdoor activities, especially hunting, have long traditions and are important elements many people's lifestyles. Most people in Våler are probably in contact with hunters and hunting dogs, and moose season is an important period in the community calendar, just as in Østerdalen. Yet, people living in densely populated cultural landscapes may perhaps react more strongly to wolves outside the kitchen window than people in the deep forests, who, after all, are used to the sight of large carnivores. Another clear difference between Østfold and Østerdalen is the former's even smaller number of rough-grazing sheep. In 2000 (the heyday of the Moss pack), according to compensation records, wolves killed only nineteen sheep and one heifer in all of Østfold. This was due probably in part to high electric fences (which the government have supported) but mainly to the low number of livestock.

Just before we started our fieldwork in Våler, a wolf and a dog were suspected of breeding a litter of hybrids. A young wolf had been run over in Våler in October 1999, and its DNA was analyzed. The test confirmed people's suspicions, and in February 2000, permits were issued to fell four animals. Responsibility was delegated to the office of the County Governor. The actual killing was carried out by a hand-picked hunting party, with several police officers instead of local hunters (in itself a source of conflict). In 2000, a new litter of seven "real" wolf cubs was confirmed, and these wolves made up the Moss pack in our study period, winter and spring 2001.

The municipality of Halden lies in the southeastern corner of Østfold and shares a border with Sweden. Historically, Halden has always been a manufacturing town. It was a "true" urban municipality for many years, but after a merger with two other municipalities, Idd and Berg, in the 1960s, it came to include an abundance of forests and farmland. The total area of the municipality today is 642 square kilometers. The population has grown steadily over the past fifteen years, and in 2010 the municipality had 28,776 inhabitants (a fair-sized regional town in Norway). Most, 85 percent, live in and around the town itself. Although most of our informants are from Idd and Berg, the urban population is also represented in our study. Historically, manufacturing emerged here because of a small but powerful river, the Tista, a part of the Halden watershed. There were once manufacturers of all varieties here, including wood processors, cotton mills, and the enamel works Cathrineholm. Halden was also known for many years as the "shoe town." Nearly every third worker still works in manufacturing, which is 10 percentage points higher than the national average. Halden still has 2,500 jobs in manufacturing. The R&D and IT sectors employ about 1,500 people.

Today, the forests and cultural landscape mostly cater to the townspeople's recreational needs. Nevertheless, local authorities stress on the municipal website the economic importance of forestry and agriculture. Halden is now the largest forestry municipality in Østfold and the second largest farming municipality, yet only 1.8 percent of the workforce is in the primary industries. There is little to be found of the farming most affected by carnivores, that is, sheep put out to graze on open, uncultivated land. Hunting is an important activity in the rural parts of the municipality, even though the percentage of hunters relative to the population cannot compete with, for instance, the percentage in Trysil. In Halden, 993 people paid the hunting fee in 2009, and the season resulted in 141 felled moose within the municipal borders; here too we note the contrast with Trysil's tally of 1,143 felled moose. Halden is the Norwegian municipality with the longest-lasting stable wolf presence. The Dals Ed–Halden pack (whose territory includes areas in Dals Ed in Sweden) has been around almost as long as wolves have been in Norway in modern times. Linnekleppen and Kynnefjell are two recently established packs near Halden. In addition to wolves, a few family groups of lynx live within Halden's borders.

QUALITATIVE METHOD AND REPRESENTATIVITY

A common methodological approach to the study of attitudes and opinions is, put simply, to talk to people. In small, local communities, qualitative methods such as in-depth interviews with individuals or focus group sessions quickly give the researcher an impression of the range and variety of opinions on a given subject. Since these methods allow participants to talk about several topics, researchers can use the opportunity to elicit the wider social and cultural context. By letting informants explain their opinions about carnivores and how the presence of carnivores affects their lives and the local community, we learn something about what informs people's different opinions. Since interviewees also talk about many other aspects of their lives and their view of the local community—past, present, and future—we can also begin to understand the context in which these views arise.

We have studied wolf conflicts for more than fifteen years. Several hundred people have participated as informants in the different research projects. We have organized many focus groups, and many people have taken part in in-depth interviews. Because we have visited some local communities so often, we have also gathered a good deal of informal data from various conversations and observations. Furthermore, from 2000 to 2002, we conducted what could be called fieldwork among young hunters. We spent a lot of time with them and participated in many of their activities.

Qualitative interviews are good for eliciting the context in which the opinions we want to learn about are formed, but they do not guarantee representativity. We attempted to compensate for this in two ways. The methodological concept of "saturation" (Bertaux and Thompson 1997) is based on the presumption that social phenomena such as opinions, attitudes, and motives for choice of action are shared by several people. There are many more people than there are opinions about things. We can exploit this methodologically. As the interviews progress, opinions begin to recur. When new interviews cease to provide new information, we can say saturation has been reached, at which point we can be reasonably sure that we have covered the range of opinions on a given subject, such as wolves. We will then discover that people's opinions can be grouped into a few main types that we will analyze and contextualize.

We have also attempted to compensate for the lack of statistical representativity by what we can call "quasi-representativity." We recruit informants that in sum could reasonably be expected to represent much of the variation in opinions. First, we approached people expected to feel affected by the carnivore management regime in a direct way, such as landowners, hunters, members of environmental NGOs, and users of pastures and other natural resources. Some informants were selected more "randomly," in workplaces, at schools, and in neighborhoods areas. These did not represent well-defined interest groups, and our purpose was to reach people who just "happened" to live in areas with large carnivores. Nevertheless, some of them held strong opinions both for and against the wolf, and many had more detached views of the subject. We also made a consistent attempt to speak to men and women, young and old, with different occupational and educational backgrounds.

The material for this book was collected in part from focus groups. A focus group session unfolds as a group conversation. One or two researchers led our conversations, which lasted between two and three hours. The degree to which the researchers controlled the interviews varied widely. Experience of and views about wolves and other carnivores readily captured the attention of participants in most of the groups. Our material includes many examples of conversations that flowed easily, with a lot of enthusiasm from the participants. Wherever we went, people with local knowledge agreed to help us recruit participants for the focus groups, spread information about the project, and find us a place to hold the conversation.

There are two main strategies when it comes to composing focus groups. In heterogeneous groups, one can—ideally—study what happens when people with different backgrounds and views discuss a subject and what happens when different viewpoints clash against one another. This, however, is not easy to accomplish. The emerging group dynamic may give one or two participants a dominant position in the conversation, leading to the suppression of other opinions. We chose instead to create more or less homogeneous groups, which consisted of, for example, members of an organization, colleagues from the same workplace, or members of the same hunting party. This method is unlikely to give us a great breadth of opinion in each conversation, but the conversations will often be quite energized because people do not agree about everything, and things that might have remained undetected in interviews with individuals can surface.

The advantage of interviewing people separately, which we have done many times, lies in the avoidance of group pressure and in allowing informants more time to elaborate and contextualize their story. Exploring a particular subject in much more detail is possible with a single individual. Since you forfeit some of the dynamics of the focus group conversations, so important social dimensions underlying the formation of opinions remain hidden, to combine the two methods makes sense. Focus groups offer, moreover, a practical and affordable way of speaking with a relatively large number of people. To arrange separate interviews with each person to obtain the same information would be practically and financially impossible.

To be able to claim saturation, we need to know something about the social system being studied. In a project like ours, this means not leaving out whole segments of the population. We therefore started each study by obtaining an overview of the communities where we were planning research. We looked at the economic structure, civic activity, and the public's use of the natural environment, among other things. As far as possible, we recruited informants from different sections of the community. When we add the temporal factor—we have been doing this for quite a few years—we feel confident that we have an overview of the diversity in attitudes towards carnivores. When we look at all our interviews together, we can say saturation was achieved.

The book is based on data collected for different research projects and analyzed in different phases and for various purposes. These circumstances are reflected in the chapters of the book. Not all informants or focus groups are included in the analyses presented in any single chapter. Most chapters build on articles published in scientific journals but to a varying extent. This is explained whenever applicable. The places and informant groups involved are presented in each of the chapters.

OUR ROLE AS SCIENTISTS

One of the first observations we made when interviewing people skeptical of wolves was that they had very little confidence in science. Many saw scientists as part of a large and powerful alliance that had engineered the wolves' return to Norway. Over the years, we have talked to hundreds of people about wolves, but we have hardly ever been met with hostility that could be ascribed to the fact that we are scientists ourselves. On the contrary, most of our informants were enthusiastic storytellers and seemed happy to have someone listening. How can that be?

In an early phase of our studies, we presented some of our findings to a group of colleagues. During a discussion on the role and status of scientists and scientific knowledge, the question came up: How was it possible for us to be accepted when science in general was met with so little acceptance? One colleague, who had lived in our study area for many years, thought she knew the answer. A friend of hers happened to be one of our informants. In a telephone conversation, our colleague had asked her friend to explain. The answer she got was, "Oh, but they are not *real* scientists." The statement was not necessarily a tribute to us or to social science in general, but the sentiment it expressed actually facilitated our data collection. People who denounced scientific knowledge about wolves saw our role as different from that of the biologists.

Even if we were looked upon as "not real scientists," we were still seen as "semi-officials" affiliated with the government system, although not in the same way as our "real scientist" counterparts. We were seen as part of the system but not as agents of the scientific knowledge that underpin wolf management. That combination opened doors to us. Many people feel it is almost impossible to make their voices heard in the face of government at all levels. Our experience was that most people did not hold back when talking to us, and several told us they saw us as potential spokespersons who could help them get their version across. Although we could promise no such thing, most informants were happy to share their thoughts and experiences, eager to explain their points of view. Only once have we asked for an interview appointment and been rejected, by a pro-carnivore NGO that did not see how their cause could benefit from talking to us. They saw our research as legitimation of rural anti-carnivore sentiments.

NOTE

1. We use the term "municipality" for the lowest administrative unit in the Norwegian government structure (*kommune*). The English term "county" is customarily used for the intermediate administrative level in Norway (*fylke*). This level also has an elected assembly (like the municipalities), but importantly it also has a strong presence of central government agencies, for example, the County Governor (Fylkesmann). The environmental authorities that operate at this level represent the national government.

CHAPTER THREE

NEW ALLIANCE, OLD ANTAGONISM

✳ ✳ ✳

The media, in Norway and elsewhere, have portrayed conflicts over large carnivores pretty much as urban-rural conflicts. This understanding also enjoys a strong position in the public debate, where a picture of unified rural communities seems to prevail. We easily get the impression that these communities stand united in their resistance against intruding vermin, urban romantics who are oblivious to the problems of rural people, and the government's heartless policy of turning once-vital rural areas into game preserves. In fact, the anti-wolf front is construed almost as a last line of defense against destructive forces threatening rural life "as we know it."

The idea that large carnivores, and particularly wolves, should be protected is claimed to be typical of urban people, who tend to have a romantic view on nature: they want it to be a wilderness they can watch on TV and that can provide scenic settings for their leisure activities. Such values are on the offensive in contemporary society, as demonstrated through the protection of more and more areas and, not least, a large carnivore management geared toward the protection of all species. This is what it may look like from a rural perspective—a perspective that is widely promoted in local media. Of course, the contrast depicted here is a drastic oversimplification. Surveys show that wolf supporters and wolf adversaries dwell in both urban and rural areas. Granted, supporters are more numerous in cities, but quite a few reside in rural areas as well, even in those with wolves (Dressel et al. 2015; Krange et al. 2017; Tangeland et al. 2010). The picture is particularly misleading when all rural people are portrayed as being against wolves. As we shall see particularly in chapter 5, our interview data supports the survey results and reveals that all typical opinions on wolves are present in the wolf areas. Therefore, the question is why the picture of a united rural front has achieved such a dominant position.

Media coverage is of course an important factor. The media generally seek clear-cut images, and if these are about conflict, even better. But the media images are definitely not without a basis in local self-presentations, which are frequently rife with explicit declarations of local solidarity and concord in the face of serious threats. Local politicians, high-profile individuals, and organizations that claim to represent rural interests often use the resistance against large carnivores for all it is worth, often presenting loss of livestock and dogs, declining game stocks, and lost hunting revenue as factors that may eventually lead to farm abandonment and the demise of whole communities. And this demise will not only be brought about by direct economic harm to farms and resource-based businesses: people will leave because of a deteriorating life quality due to diminished options for outdoor recreation and concern for their own and their children's safety. In sum, this will lead to erosion of the economic *and* social basis of the affected communities.

A conspicuous aspect of this collective self-presentation is the way it depicts carnivore supporters as a contrast to rural people. Local media and various websites usually portray carnivore supporters as urban romantics who have lost touch with the basis for human existence. The authorities that execute official large carnivore policy are said to be ruthless and without compassion for rural people, essentially geared toward accommodating urban middle-class interests. Interestingly, a new term has become a buzzword in the rural discourse on large carnivores (and other aspects of nature conservation): the Norwegian phrase *storsamfunnet*, which translates into something like "society at large." *Storsamfunnet* is frequently named the real culprit: *the* actor that (through the state) forces small rural communities to live with the predator pest and to give up traditional land use rights so urban people can have protected areas as their playgrounds. The term is often used in a deliberate manner, intended to emphasize the uneven balance of power between big institutions and urban centers on one hand and small rural communities on the other. Paradoxically, however, ordinary people living in cities are obviously part of *storsamfunnet,* while not even the elites in rural areas seem to be included.

Thus, a powerful rural discourse seems to invoke an image of communities under dangerous pressure from hostile external forces. However, "communities" do not grow out of the earth in any given locality, rural or otherwise; they are actively constructed. Given that any spatially delimited subsection of society is divided along lines of class, education, gender, generation, and so on, we must, in a sense, *choose* to see the people who live in a particular place as a community. We will not treat this as a conscious process at the individual level but rather concentrate on the collective aspects of the construction of community.

In rural areas, social diversity has increased throughout recent decades. General processes of modernization have made rural social systems more complex than they once were. The educational profile of their workforce has changed significantly. Employment in agriculture, resource extraction, and related processing industries has declined sharply, whereas middle-class jobs demanding higher education have increased. This change is due to an expanding public sector and a shift toward a service sector economy, which has been quite significant in some rural areas. As demonstrated by research in the United Kingdom and the United States, social diversification has also increased through in-migration of middle-class people who are not part of the local labor market but either work from home or commute to urban centers (Bell 1994; Nelson 2001). Semi-urban centers have emerged, mirroring as well as driving economic, social, and cultural change. Exchange with larger urban centers is extensive, and the Internet has certainly not passed rural areas by.

Of course, rural areas were never culturally homogenous, as very diverse living conditions were reflected in different understandings of the local social structures and of the world in general. Different cultural patterns have always reflected material class differences and interest conflicts, not least related to property rights. Even so, the rural sociocultural landscape has become more complex over the past few decades, which is perhaps particularly expressed among younger people (see Farrugia 2013; Skogen 2001). Accordingly, and not least due to weakened ties to natural resource utilization, attitudes toward nature and land use issues can hardly remain unaffected. That such attitudes are common to all people living in rural areas is unthinkable. General population surveys provide us with an example: many who live in areas with large carnivores want to have these animals around, including wolves.

Obviously, then, strong community images invoked through the wolf controversies do not simply reflect common material interests or a monolithic culture. The task here is to explore some theoretical contributions to the concept of community in order to see how they might help us understand the wolf conflicts. However, while the socially constructed community invokes a picture of unity, it can serve different functions for different groups (see Cohen 1985; Liepins 2000). We shall now take a closer look at contributions from the social sciences that may help us understand this particular dimension of the conflicts over large carnivores.

THE SOCIAL CONSTRUCTION OF COMMUNITY

"Community" does not have one single meaning but many. The term has been part of different research traditions throughout the twentieth century but is also used in different ways in contemporary strands of social theory. However, the most common one may be what Ruth Liepins (2000: 25) calls a "minimalist approach," meaning that no real definition is provided but that the term is used to denote a "loosely specified sense of social collectivity." Many "community studies" are studies of what we may term "local social relations" (Crow and Allen 1994) that do not necessarily need a more stringent definition. Many aspects of spatially delimited social systems, rural as well as urban, can be analyzed without any sophisticated definition of "community"—indeed, without relating to the term at all. However, community *as* perceived collectivity is also an important research issue. One of the most publicly visible accounts of wolf resistance is that it is a unifying force in rural communities, underpinning the strong sense of collectivity people in such areas feel. The emphasis on external threats, particularly from *storsamfunnet* ("society at large"), and on internal coherence draws our attention to the issue of community boundaries. How are these boundaries constructed and managed, and by whom?

In our view, the most productive theorizing on community boundaries is found within the "symbolic construction approach" (Liepins 2000), where the standard reference work still seems to be Anthony Cohen's *The Symbolic Construction of Community* (1985). And with good reason: Cohen persuasively argues that communities as collectivities can have no objective existence. They are not only social constructions but also contested and furnished with very different meanings by different members of any given community. As a basis for our further discussion of the carnivore conflict as an element in the construction of rural communities, we will outline some staple elements in Cohen's theorizing. A simple yet fundamental argument in Cohen's work is that the concept of community implies both similarity and difference. Those inside share something that makes them different from those outside. Therefore, community is a relational concept, inconceivable except in relation to something different from it. However, the internal similarities—common values, beliefs, norms—that constitute the inside of the community are not so definite. Quite the contrary, these "big" categories tend to be rather elusive, and their content is almost impossible to define with any precision (Cohen 1985: 15). And given the economic, social, and cultural diversity within any community, a total congruence of the interpretations of such concepts is hard to imagine. But community as a social construction is a symbol, and symbols do not simply convey meaning; they are tools we may use to produce it. If symbols had a very exact content, we would not need them. Indeed, the flexibility of symbols may be their most important quality. So, within limits, the same symbol may express different meanings for different people, while still emphasizing internal unity. Diverging interpretations of the same symbols (symbolic expressions of the community) do not hinder an effective construction of community. On the contrary, such a divergence is a prerequisite: a community could not be upheld as a social construction if people could not use the same symbol to express different perspectives, indeed to pursue their own interests (economic or otherwise).

The idea of community demands that "inside" and "outside" are clearly separated, which draws attention to the place where the two spheres meet, that is, to the boundary. Boundaries do not simply exist; they must be constructed. In some cases, they may be easy to see, even for outsiders, but in others they may be almost imperceptible to all except the members of the community itself. Since the boundaries are social constructions, they must be demarcated symbolically. Community itself is a symbol that fills this function of upholding boundaries. As Cohen (1985: 15) puts it: "Community is a boundary-expressing symbol. As a symbol, it is held in common by its members; but its meaning varies with its members' unique orientation to it." And further: "The triumph of community is to so contain this [internal] variety that its inherent discordance does not subvert the apparent coherence which is expressed by its boundaries."

The issue of constructing boundaries is also a central theme in the writings of Mary Douglas (Douglas 1992; Douglas and Wildavsky 1982). Her focus on danger (or risk) and blame in this regard seems relevant for our analysis of the carnivore conflict and its function in the symbolic construction of community. Placing the blame for dangers that threaten a group on those thought to deserve it is an important element in the construction of the boundary between the group and its surroundings. Which dangers we choose to worry about—among all those we *could* worry about—are generally not based on probability calculations or knowledge of "facts," nor is the choice arbitrary. Rather, dangers that pose the most critical threats to central values or "moral principles" are generally taken most seriously. And placing the blame for this type of danger on actors outside the community is important to enhance internal cohesion and thus also to construct visible boundaries. Douglas and Aaron Wildavsky (1982) take the environmental movement as an example of a "community" that aims to achieve internal cohesiveness by positing big corporations and their allies in the state apparatus as an outside enemy, which is demonized and held up as everything the environmental movement is not. Ironically, the environmental movement itself is often pictured as part of the evil conglomerate confronting people in areas with large carnivores and in this context is attributed with many of the same qualities as the environmental movement ascribes to big business and the state. Together with management agencies and biologists, the environmental movement is blamed for posing unacceptable danger to the community (jeopardizing the safety of people and domestic animals and not least the experienced "life quality"). Together with ignorant city romantics in general, these actors are held responsible for the perils many rural people now face.

Rural communities are in the midst of significant social change. Economic modernization, cultural diversification, and increased social and spatial mobility weaken the basis of traditional rural communities built around agriculture and resource extraction. Cohen (1985: 70) writes, "When the structural base of the boundary becomes undermined or weakened as a consequence of social change, so people resort increasingly to symbolic behavior to reconstitute the boundary." Why do they do this? Obviously, some changes actually threaten the interests of individuals or groups, and resorting to collective resistance efforts is a "rational choice." Additionally, and affecting even more people, these structural and cultural changes pose a threat to individual and collective identities. The concept of community is inseparably tied to the concept of identity. As Akhil Gupta and James Ferguson (1997: 13) write, "'Community' is never simply the recognition of cultural similarity or social contiguity but a categorical identity that is premised on various forms of exclusion and constructions of otherness." Thus, symbolically reinforcing the community and its boundaries will constitute a resistance effort in the face of threatened individual and collective identities. This resistance may, however, have a different content for different groups in "the resistance front" within the community.

The local sense of community has generally presented itself as a core issue in or studies of the wolf conflicts. We observed this phenomenon very early during our fieldwork in Stor-Elvdal in 2000, which is why the following analysis is based mainly on material from that study site. But we have come across the same in many places, and we can safely say the picture we will be drawing here has a more general validity.

PEOPLE AND WOLVES IN STOR-ELVDAL

When we interviewed people in Stor-Elvdal, two wolf packs lived within the municipality's borders. The inhabitants already had much experience as the neighbors of wolves. At this time, Stor-Elvdal received a good deal of attention from the national media, particularly since authorities decided to eliminate one of the packs because it was in one of the areas within the region that actually has some sheep farming. This caused considerable media buzz, and several locals were interviewed in newspapers, radio, and television. However, the media were not interested in talking to just anybody. Most often, they let ardent wolf opponents talk about their views because they were taken to represent the characteristic "rural" view of wolves. This was done not to help rural people to get their message across but rather, at least as understood by the locals in hindsight, to portray them as backward and bloodthirsty. We will return to this particular experience in chapter 4.

We soon found out that opinions varied more than the media portrayed. All typical views of the issue were present, including very strong pro-wolf attitudes and critical views of modern sheep husbandry and of common hunting practices. With very few exceptions, however, people holding such views kept a low profile. We think this normally happened not because they feared reactions (although that may have also been a consideration for some) but rather because the issue was not as important to them as to the wolf adversaries (though outspoken wolf supporters were also present). A sizable group was also not engaged in the issue at all. Our data do not allow us to determine how large parts of the population tilted one way or the other in the wolf question, but the impression from all our study areas, including Stor-Elvdal, has been that considerable groups fall into the "not interested" category. Nevertheless, various degrees of anti-wolf sentiments were common, which, as discussed in the introduction, has strongly influenced our perspective in this book.

We found that varieties of wolf opposition could be grouped along two axes, one economic/practical and the other cultural. These axes touch each other, but they are far from identical. Three principal groups were identified: sheep farmers, landowners who lease hunting, and local people with strong ties to traditional land use practices (primarily hunting) and roots in a rural working-class culture. These groups have not always been allies. In fact, conflicts of interest are easily identified, and local history is rife with class antagonism. Scratching the surface was often enough to show us that this history was very much alive. We were told more than a few stories about the strained and often openly hostile relationship between rich landowners and the local working class that once prevailed. Current interest conflicts were also visible when we did our fieldwork in Stor-Elvdal.

THE LANDOWNERS

For decades, moose hunting has been of a certain economic importance for owners of large forest properties in this region, although of course completely overshadowed by timber sales, the basis of tremendous wealth for a few families. Lately, however, some of the largest properties have developed hunting tourism into a more important part of their enterprise, as a response to uncertain times in the timber industry and a general trend toward commercialization of outdoor recreation activities, particularly hunting and angling. But just when the landowners have established big game hunting as a significant part of their business, enter the wolf, which eats moose and kills hunting dogs. To what degree moose stocks will actually decline over time as a result of wolf depredation is subject to debate. However, there is no doubt that wolves have an impact in the areas where they concentrate their hunting at a given time (Gervasi et al. 2012). Likewise, wolves do indeed attack and kill dogs. There have been many episodes in Norway and even more in Sweden. Some dogs have been killed in Stor-Elvdal too. The landowners feared that the presence of wolves would severely reduce the value of their hunting. There were already reports of hunters who had turned their backs on landowners whose hunting grounds were invaded by wolves.

The region where Stor-Elvdal is located has some of the largest private forest properties in Norway. The owners belong to an upper class where higher education and cultural capital is abundant. They generally expressed their skepticism toward the wolves in a civil manner, often referring to research and current public debate. Understandably, they focused on the economic losses they suffered and on the negative impact the wolves could have on developing hunting tourism. Some of them seemed to conclude that the wolves would have to be removed from Stor-Elvdal (although they stressed that wolves should definitely be protected in "real" wilderness areas). Others were more pragmatic and suggested that changes in the management model and economic compensation (for lost hunting revenue) could be sufficient. A few had seriously considered cooperating with the World Wildlife Fund (WWF) and other outside actors to arrange wolf safaris but had decided against it, seeing it partly as a bad tactical move and partly as an act of betrayal against fellow landowners and other suffering community members.

They all agreed the state should compensate their losses. At the time of our fieldwork, a group from Stor-Elvdal had actually persuaded the Ministry of Environment to try out an arrangement in which landowners with "established wolf packs" actually received compensation. While highly controversial, as many saw it as a step toward establishing property rights to wild animals, the arrangement also provoked smaller landowners in the region who considered it a form of prostitution. However, many landowners may see such an arrangement as the type of economic and practical solution that they can live with and which they have adequate resources to actually bring about. The project was dropped after a few years. A new government came in, the controversy did not subside, and a permanent and general arrangement would have been extremely expensive and difficult to manage.

THE SHEEP FARMERS

Sheep husbandry was not a common activity in Stor-Elvdal until the 1970s. It only became economically feasible through subsidies introduced at that time to bolster farming in marginal areas. In a situation without large carnivores, authorities saw rough grazing of sheep as one of few viable productions that could be established in regions not suitable for large-scale farming. The program has been a success in the sense that this form of sheep farming has become widespread, and the number of sheep in Norwegian forests and mountains has increased dramatically since the 1960s. However, the region in which Stor-Elvdal is situated does not have many sheep compared to areas further to the north and west. Around 2000, Stor-Elvdal had about thirty-five farms with sheep, only six or seven of which had sheep as their main activity, meaning sheep farming did not contribute much to the local economy and involved few people. Furthermore, sheep breeders in Stor-Elvdal were a mixed group. Some of them engaged in a diversity of economic activities related to land use and tourism, some were large landowners with sheepherding as a minor sideline, some were well-educated people who had decided to leave the city behind, and of course some were traditional farmers.

The sheep owners' problem with the carnivores is easy to see: their sheep are eaten. Although many consider the economic compensation generous, killed sheep are a big problem for the farmers. It causes extra work and represents emotional strain that is easy to identify with. The sheep owners' situation has received considerable attention in the national media and in public debate, not least because these challenges are quite tangible and graphic, and they may stir sympathetic feelings even in distant media consumers. Authorities have also concentrated on the problems the carnivores cause sheepherders, since they relate to economic activities, including those that receive government subsidies. Furthermore, farmers are represented by strong organizations that government agencies are accustomed to dealing with, and farming is still a backbone in Norwegian regional policy, aimed at maintaining settlement all over the country. This has contributed to defining problems caused by large carnivore as farming problems, not least on the national political arena.

Following the arrival of wolves in Stor-Elvdal, sheep farming received more attention than ever before. Some local sheep owners were resourceful people who often managed to make their voices heard. Now they were also helped by the media focus on their problems, not only depicting them as over-subsidized receivers of taxpayers' money. Sheep owners could now see themselves as symbols of rural people's struggle against urban ignorance, a role in which many people outside sheep farming also saw them. Very few sheepherders could accept wolves in their area, or in the whole of Norway. Their arguments generally focused on rough grazing as a viable, ecologically sound method of meat production, which, as the wolf is not threatened as a species globally, should be considered more important than protecting the wolf in areas where such production takes place. However, the way this view was articulated varied a great deal. We might say that one extreme was an economic rationalism akin to the perspectives of some forest owners, while the other was a form of "cultural resistance" (a concept we will discuss at some length in chapter 4).

THE HUNTERS

A significant portion of wolf adversaries was found among working-class men with a strong attachment to a traditional lifestyle close to nature, particularly through hunting. The concrete issues these men focused on were the loss of hunting dogs and the decline in some game stocks, primarily moose and roe deer. The presence of wolves forced them to hunt less and in other places. Knowing the affectionate relationship between hunters and their dogs, as well as the tremendous amount of time (and money) many hunters invest in training dogs, it was no surprise the wolves were not popular. Indeed, the typical Scandinavian hunting methods, which entail the use of untethered dogs, were now seen as impossible in areas with wolves. Since many hunters regard cooperation with the dog as more important and rewarding than the actual kill (Krange and Skogen 2007b), the loss of this form of hunting was all the more aggravating.

These men were strongly attached to the place where they lived and to the land. In several respects they kept up a traditional way of living typical for men in rural areas, which entails largely manual work and a somewhat rough contact with nature. The men were firmly rooted in what we might call a production-oriented culture: cultural forms that have grown out of the everyday life of workers and farmers, comprising high valuation of practical work, technical ingenuity and masculine toughness, and skepticism toward academic knowledge and intellectual pursuits. The hunters were just as angry with environmentalists and wildlife biologists (roughly perceived as the same group) as at the predators themselves. They felt that city people generally had far too much power, which they were now using to turn rural Norway into a game preserve. Due to ignorance and indifference on the part of powerful organizations and government agencies based in urban areas, the voices of local people were not heard.

Underlying such views was an experience of being subject to patronizing attitudes from people who do not know Stor-Elvdal and a strong feeling that local people's knowledge about nature was not taken seriously. Seen in this perspective, the dominant discourse of carnivore protection is a typical instance of middle-class efforts to shape and correct the opinions and attitudes (indeed way of life) of working-class people[1] (Dunk 1991, 1994; Krange and Skogen 2011; Skogen and Krange 2010).

ALL THE REST

This chapter's main focus, like in much of the rest of the book, shall remain on the wolf adversaries. We want to understand the element of the social dynamics of carnivore resistance that has to do with symbolic construction of community. But it is essential to keep in mind that a considerable number of people in Stor-Elvdal did not engage in the wolf issue, and quite a few welcomed the wolves. Such has been the case in all of our local studies and has been confirmed in national surveys. Although the following discussion will concentrate on the wolf opposition, we will eventually see that these less dedicated groups did not escape the symbolic construction of community.

A SENSE OF COMMUNITY

Despite the differences between them, sheep farmers, hunters, and landowners all told stories not only about the practical problems the predators caused but also about general issues, such as declining quality of life due to people's fear of wolves. They frequently claimed that a principal asset of life in rural areas, namely outdoor recreation, was seriously devalued because many people are afraid to go for walks, especially to take their dogs out. There was a particular concern for small children who were allegedly not allowed to play outdoors alone anymore. Elderly people, especially women, were seen as another strongly affected group. By picturing "weak groups" as vulnerable to dangers imposed on the community from outside, they emphasized that wolf protection was cruel and inhuman, representing an infringement on the community and in effect an assault on the "weakest among us."

Furthermore, problems that mainly affect certain groups were frequently presented as serious concerns for the community as a whole. This was done not only by those who experienced a particular problem but other groups as well. For example, the hunters whose chief concerns were (perceived) dwindling game stocks and attacks on dogs often pointed out the problems experienced by landowners who leased the hunting the hunters had to pay for. And sheep owners, a formerly anonymous group that few people paid attention to, were now heralded as vanguard defenders of rural lifestyles and the very inhabitation of marginal areas.

CONTESTED KNOWLEDGE

Rival knowledge systems play an important part in shaping the carnivore conflicts (more on this in chapter 6). Schisms originating from the tension between scientific knowledge and lay, tacit knowledge will overlap and merge with conflicts stemming from tensions between hegemonic and subordinate cultural forms. Such overlap is not perfect, but hegemonic cultural forms generally coincide with or subsume scientific forms of knowledge, as opposed to subordinate cultural forms, where practically founded knowledge holds a pivotal position. When government policies, such as those that underpin large carnivore management are based mainly on input from certified experts, many people feel that practical, lay knowledge is not taken seriously. Abstract scientific knowledge enjoys a dominant position within institutions of power.

In social segments with strong ties to a production-oriented culture, practical, experience-based knowledge, and technical ingenuity are cultivated and seen as superior to abstract, airy-fairy "desk knowledge" (see chapters 4 and 6; Krange and Skogen 2007b, 2011; Willis 1977, 1979). Working-class hunters constitute one group that puts up cultural resistance through actively defending practical experience and ridiculing academic knowledge and academic pursuits (Krange and Skogen 2007b, 2011; Willis 1977, 1979), represented not least by official biological knowledge about large carnivores. Wildlife biologists and managers were not only considered wrong; they were also accused of dishonesty and manipulation, for example, regarding population numbers. Skepticism and ridicule related to abstract, academic knowledge are features portrayed in many studies of working-class culture, as we will discuss in more depth in chapters 4 and 6.

The present analysis expands this picture, as *all* groups of carnivore skeptics claimed that local knowledge was generally snubbed by those in power, whether they were politicians, managers, biologists or conservationists. However, this was expressed in a relatively subdued manner by landowners, who were often well educated and thus more loyal to scientific discourse. The same applied to quite a few sheepherders, who, as we have seen, were not always traditional local farmers. These people expressed a considerable ambivalence toward scientific knowledge, for example, regarding the sizes of carnivore populations. For them to show that they were in no way ignorant was obviously important, while at the same time they defended local and practical knowledge against what they, too, perceived as disparagement and often ridicule. Thus, they joined the working-class hunters in constructing a boundary around the local community by letting antagonistic forms of knowledge demarcate the line between "inside" and "outside." Local knowledge was portrayed as common to the people who make up the community and fundamentally different from the hegemonic external knowledge that legitimized the perceived assaults on rural livelihoods and "the rural way of life."

A THREATENED LIFE-FORM

Which common threats do the groups that make up the anti-carnivore alliance face? In a sense they find themselves in the same boat, albeit in very different ways, as people who stand to lose from urban expansion and related economic and cultural changes. We also see here an example of cultural commonality between the working class and the property-owning upper class: the defense of material production—and associated values—against the cultural expansion of the modern middle class. This expansion entails, among other undesirable things, extensive nature protection based on a romantic view: nature seen as delicate and vulnerable, always threatened by human activities (i.e., the activities pursued by the working class, farmers, and landowners). Also, the new middle class may be seen as the culprit behind the mass of regulations interfering with every conceivable aspect of human existence, not least private enterprise, and here elements of working-class and "bourgeois" culture tend to converge (Skogen and Krange 2010).

Landowners, farmers, and working-class hunters all talked about "our" way of life as threatened by current carnivore management. They often said the presence of wolves seriously disturbs the ways "we" use the land, which was clearly not a simple reflection of common lifestyles and land use practices, as these were rather different. Furthermore, the ways these groups use the land may indeed come in conflict with each other, as when hunting dogs chase sheep and sheep farmers with dogs disturb hunting in the fall. To some extent, they even reflect antagonistic economic interests, as is the case with landowners who want to maximize their profits from hunting and the local working-class hunters who have to pay for it. There were similar conflicting interests crisscrossing the social landscape between all three groups.

However, cultivation of the rural way of life as a defense against urban expansion appeared to be a common identity factor, despite cultural and economic differences. This rural lifestyle was constructed in different ways, as exemplified by the contrast between oilskin raincoats and expensive SUVs on one hand and baseball caps and old pickup trucks on the other. All varieties of rural lifestyle may be exposed to threats due to current processes of social change, although lifestyles based in different class positions will not be affected in the same way—or necessarily by the same aspects of social change. The appearance of wolves may then be seen as a symbol of changing value orientations in "society at large," which changes may be tied to the more tangible changes rural people experience in their everyday lives.

Michael M. Bell (1994) provides a very convincing account of "construction of community" in an English village with substantial in-migration of wealthy urbanites and a population generally characterized by both old and new class antagonisms. Among Bell's important insights, we find an explication of the role of nature as substitute for class as a source of identity—indeed, as a source of morality. Nature escapes the problematic aspects of class that trouble Bell's informants: although they recognize the continued significance of class in most areas of life (irrespective of their own social position), they claim this is a deplorable state of affairs. By seeking refuge in the concepts of "nature" and indeed "rurality" as significant elements in their identity projects, they are able to construct a sense of collectivity that, at least at one symbolic level, overrides the blatant material inequality and large cultural differences that otherwise mark the village. Seeking and defending nature, as well as emphasizing a fundamental antagonism between rural and urban life, will let wealthy urban in-migrants and the old "landed gentry" slip away from confrontations over their material privileges and political power and may simultaneously provide the rural working class with components for identity construction that downplay their material and political deprivation. This perspective would seem to fit our observations in Stor-Elvdal almost perfectly, although, as we shall see later, the issue of class is perhaps swept under the rug even more effectively than in Bell's English village.

Varieties of this interpretation were most prominent among the hunters, as we shall return to in chapter 4, but were also found in the two other groups of wolf skeptics. One should perhaps think that farmers would be among the most anti-urban, but as we have seen, not all Stor-Elvdal sheep owners were typical farmers. Some were, and these people held views corresponding to what we describe above. However, some sheep owners had an urban background, a higher education, or both, which appeared to modify but not eradicate the anti-urban sentiments. Similarly, many of the forest owners held university degrees and also had strong ties to cities—that is, not only through their economic activities or networks from the university days; they generally had close relatives living in cities. In fact, urban speculators took over most of the large forest properties in the entire region from the 1870s up to around 1920, due to a wave of bankruptcies that drove the old proprietors away. Some of the incoming families did not settle permanently in the valley until the 1920s or even later. Thus, few of the families that own the huge forest properties have more than a couple of generations in the area, and most retain big networks in cities—networks that are generally also economically involved in the properties.

But even so, sheep farmers and landowners joined the working-class hunters in their lamentation of cities and urban life and in their conception of urban expansion—physically, economically, and culturally—as a threat to a rural life-form perceived as "common" to them all. The parallels to Bell's (1994) account form the British countryside are striking: not only will this "unification" and boundary demarcation let them confront "the enemy" in a more efficient manner; it will also take their own minds off the troubling schisms between them, tensions that from time to time break out in disturbing forms of open conflict.

AMBIVALENCE AND CLASS CONFLICT

Ambivalence seems to be a theme running through the "alliance" and underpinning the flexible way its basic perspectives are handled. As we have already seen, well-educated people negatively affected by the wolves struggle to maintain a delicate balance between urban modernity and academia on one side and a rural life-form and identification with deep-rooted local knowledge on the other. Another form of ambivalence evident among sheep farmers, landowners and hunters alike is inherent in the mixed feelings most informants seemed to have toward other groups within the alliance. These (manifest or latent) conflict lines converged with class boundaries to a significant but by no means full extent. For example, although we have focused on local hunters rooted in a working-class culture, not all hunters belonged to the working class. Practically all of the big landowners were hunters too, and in that capacity, they saw the world through hunters' eyes. But hunting was not their primary interest or source of identity.

Some sheep farmers told stories about how agreements on grazing had been discarded almost without notice because landowners wanted to give priority to wild game in order to earn more money from hunting or even because they wanted to take up a little sheep breeding themselves—possibly more as a hobby than anything else. In general, the small sheep farmers seemed uneasy with depending on the larger landowners' goodwill, as some of them had to. Some had formal grazing rights on property that was not their own, but those who depended on informal arrangements were in a vulnerable position, obviously reminding them of a time when class antagonism was more blatant. Some landowners admitted they were skeptical of sheepherding on the present scale, or at least until the wolves arrived, mainly because of conflicts with hunting—not only their own but also of people they wanted to see paying to hunt on their property. Some mentioned competition between sheep and wild game for grazing resources, but this was not seen as the main problem. What bothered the landowners was the massive presence of the sheep themselves, as well as sheepdogs and people looking after sheep, in the hunting season. This disturbs the game, distracts the hunting dogs, and necessitates extra safety precautions from the hunters. Consequently, some informants openly said they saw the sheep owners principally as tactical allies.

Local hunters are obviously vulnerable to further commercialization of hunting, and here lies a tangible conflict of interest vis-à-vis the landowners (Øian and Skogen 2015). Access to small-game hunting is still ample (and cheap) in the region, and even moose hunting is available to most locals, even though some we talked to saw the prices as rather steep. At the time of our study, there was a strong political pressure on landowners to commercialize and diversify, so as to reduce their dependence on traditional resource use (mainly timber). Accordingly, some large forest owners displayed a growing market orientation. Although there is still enough hunting for everybody, some of our hunters were seriously worried about the outlook for the future. They were well aware of the development, and some did not hesitate to label the landowners greedy and selfish. The landowners' alleged lack of engagement in local economic development was also mentioned in this connection, as an illustration of their deficient social conscience.

Apparently, there had never been much friction between hunters and sheep owners in Stor-Elvdal, something that may perhaps be attributed to the relatively low density of sheep. From other parts of Hedmark we know that hunters have clashed with sheep owners over several issues, one being hunting dogs chasing sheep and another the practice of letting sheep graze near roads and railways. When these sheep are inevitably run over, hunters who serve on municipal wildlife control teams are dispatched to put them out of their misery, which is not seen as responsible animal husbandry. However, the Stor-Elvdal hunters had not been very aware of the sheep owners and their problems until the wolves arrived. Bears had caused problems for livestock herders in the region much longer than the wolves and actually forced some sheep farmers to give up rough grazing before a single wolf was spotted. But the bears are not a problem for hunters, and some informants admitted they had only recently discovered the true anguish of the sheep owners but claimed they would never again forget it, even if the wolves should disappear.

In general, there seemed to be limited contact between the three groups; we might say there was limited contact across class boundaries. There were, however, some notable exceptions. One was a particular type of relationship between a few working-class hunters and "public-minded" landowners (see also chapter 4), where the former were recognized as expert hunters and received different types of favors in return for helping with hunting-related tasks. They also did odd bits of work for the landowners, for example, repairing cabins or building hunting platforms. While these relationships seemed to be genuinely cordial, they nevertheless had "master-serf" written all over them. Another example is the contact between landowners and sheep owners in the former's capacity as farmers and in the latter's capacity as foresters. Although farming meant very little economically to the larger landowners, some of them did farm. Some sheep owners also owned forest properties, although they were generally small and sometimes not very productive. Thus, there was a farming-and-forestry community of sorts, associated with farming and forestry organizations, the promotion of farming and forestry interests vis-à-vis local authorities, and so on. However, social contact seemed limited and in some cases molded in the same asymmetrical form as the relationship between working-class hunters and big landowners.

Bell (1994) found that class was an omnipresent yet disturbing factor in his English village's everyday life. In Stor-Elvdal, the existence of class boundaries was largely rejected by all of our wolf adversaries (and by many other informants in the study; see chapter 4 for discussion of new types of class boundaries). Interestingly, the absence of cultural, not economic, class differences was emphasized and held up as a contrast to "earlier times." Everybody, including the landowning local bourgeoisie, talked about the huge differences and impenetrable boundaries that once existed. People commonly claimed this was all gone and that everybody was now on equal footing. Quite a few of our working-class informants spoke of the younger generations of landowners as ordinary people, just like anybody else. They coached children's football teams, chatted cordially in local stores, and invited people into their kitchen to discuss hunting. The fact that they owned huge properties, lived in big mansions, and drove Jaguars was seen as irrelevant and unimportant in the big picture, and it was rarely commented on. So even though the inequality in wealth prevailed, the important thing was apparently that the landowners had changed their style. In the same vein, some landowners explained to us how difficult it once was for rich kids to be accepted by the majority of schoolmates who were "regular folk" and how tough it was to be socially isolated as a child. But the prevailing narrative was that all this had changed: there were no barriers between children from different backgrounds anymore.

The important thing here is not that this picture is quite different from the impression an observer from outside will get, namely that of an exceptionally visible class structure. Rather, we should notice there is a common propensity to construct such a vision of community, by downplaying contrasts obvious to not only visiting researchers but also many in-migrants in Stor-Elvdal. Maybe the common resort to nature and rurality as sources of community, or collective identity, has been even more efficient here than in Bell's village. Some cultural differences between Norway and England regarding the acceptability of the word "class" are possible: to recognize class as an organizing principle could be seen as more disturbing in purportedly egalitarian Norway than in Britain, which is generally seen, also by Britons themselves, as a "class society" (Bell 1994). However, it is important to bear in mind that the loggers and mill workers in Østerdalen have traditionally been among the most militant in the Norwegian labor movement. The Norwegian Communist Party had its stronghold here longer than anywhere else, and in one of Stor-Elvdal's neighboring municipalities it had two representatives in the assembly up until 2007. So "class" has not been absent from local discourses for a long time.

MORE AMBIVALENCE: A BASIS FOR ENLISTING THE NEUTRALS

If we return briefly to the people of Stor-Elvdal who had a relatively positive attitude toward carnivores, including wolves, it is interesting to observe that many of them viewed current management of land and wildlife, and indeed of large carnivores, as centralized and far removed from the communities affected by management actions. Even highly educated middle-class people who generally sympathized with conservation and who could see the wolves as an interesting addition to local nature often agreed that powerful actors located outside the community often ignored local knowledge and interests. They often resorted to the same rhetoric as the carnivore adversaries, particularly emphasizing the pressure from "society at large," yet without drawing the same conclusions about the animals themselves. Thus, even people with a positive view of the large predators, who may be generally supportive of current management objectives, seem quite prepared to construe management agencies and practices as malign outside forces. Similarly, they accentuated their bonds to local groups that disagreed with them on the carnivore issue but saw them primarily as fellow community members. They were prepared to take part in the symbolic construction of boundaries between "inside" and "outside" the community. If they somehow had to choose between wolves and "community," most of them would probably choose the latter and willingly take part in its defense, symbolic or otherwise.

We have seen here that different factions within the "anti-carnivore front" approach the conception of "community" from different angles, adjusting it to their own perspective. Those who lack strong opinions on wolves are drawn into the defense of the community because they largely understand the situation in the same way as the wolf adversaries. This is in line with the theoretical positions we presented at the beginning of this chapter and seems to support the assumption that symbolic construction of community as a conceptual framework is indeed helpful in understanding important aspects of the conflicts. To a considerable extent, these conflicts (their tangible material core notwithstanding) appear to entail tensions derived from processes of more general social change, and symbolic construction of community could be seen as part of a cultural defense line against danger from outside. Tensions between the different groups in the anti-carnivore alliance do not render their sense of community "faked" or "artificial"; it just means community is something that must be actively worked on, that is, constructed, and so is not simply a mechanical reflection of common material interests.

NOTES

This chapter is a revised and extended version of Skogen and Krange 2003.

1. The local hunters with their roots in a rural working-class culture and strong place attachment will receive considerable attention in the following chapters. Therefore, a caveat is in order: we are fully aware that "hunters" come in many varieties. Not all have a rural working-class background, and not all are against wolves. However, the hunters we encountered in the wolf areas, who denounced wolf conservation and built their identity projects largely around hunting, most often had such a background. That hunting can be a core element in rural male working-class culture has also been shown in other studies, for example, in the United States (Boglioli 2009) and Canada (Dunk 2002). We will discuss the cultural significance of hunting, for this social group, in chapter 4.

CHAPTER FOUR

HUNTERS AND WOLVES
FIELDWORK IN A RESISTANCE GROUP

❄ ❄ ❄

It was November and a late rainy evening. Through a dark landscape, we were driving along a steep and narrow forest road. There was no moonlight and no snow; the leaves had fallen from the trees. The car stopped by a stretch of marshland, covered by mist. In the distance, the silhouette of a heavily forested hill was faintly visible. When the engine was turned off, a complete silence surrounded us. Our key informant, Frank, a young local woodsman, stepped out of the car. He waited a few seconds, listening, and then he began to howl like a wolf. The vigorous, spooky sound cut through the darkness. He waited, but there was no response, and we walked silently onto the marsh. A new howl from Frank, but no response to his second or third attempts. Then suddenly, we heard something move in the bushes nearby. A rush of adrenaline, and we were running back to the safety of our car.

"It was probably nothing, but I don't trust those bastards," Frank said, referring to the wolves. Perhaps the whole thing was a con. Nevertheless, he had made me, an intellectual city guy, feel the fear. By running with him, I accepted his interpretation of the situation and demonstrated my confidence in him. After the incident, he set the terms, and the conversation went easily. Frank, who knows the woods like the back of his hand, was pointing in all directions and telling stories about his life as a devoted hunter—from childhood to young adulthood. Then he spoke of the large carnivores, especially the wolves that had recently appeared in the area. He was demonstrating what seemed like fascination, fear to some extent, but primarily anger. So, I asked him if he were not a little bit fascinated, in spite of his reasons for hating the wolves. "We can accept a limited numbers of lynx, wolverines, and even bears," he responded, "but never wolves." And as we talked about wolves, his story was transformed from pleasant memories into stories about poor prospects for the future.

(Extract from field notes)

FRANK AND HIS FRIENDS

Frank is a young man living in Stor-Elvdal, a typical forest community. Now in his late twenties, he is in transition from youth to adulthood and has recently established himself with a wife, a baby, a dog, a house, and a car. Judging from his appearance, he is the stereotype of a young rural man. He usually greets us with a firm handshake, wearing well-worn boots and denims, an old baseball cap, a green hunting jacket, short hair, and narrow whiskers. He always carries a rifle when he walks in the woods, and since he walks a great deal, he is often armed. All his life he has lived in the same small community. He has a working-class background and never completed any education exceeding the compulsory nine years. Today he is employed in the public service sector as an unskilled assistant. Over the years, Frank has seen many of his childhood peers leave the area for education or work, and he has observed that most never return. Yet he has stayed behind. He is not alone in this, however, and he has daily contact with a fairly large network of like-minded "stayers." They, too, are young working-class men who share Frank's style, passion for hunting, and attitudes toward large carnivores. Some, like Frank, have just started families. Their greatest concern these days is the reappearance of wolves in their forests.

In the latter half of the 1990s, the number of wolves increased in southeastern Norway, and some of them settled in Stor-Elvdal. These were part of a slowly recovering Swedish-Norwegian wolf population. In some segments of the community, the arrival of wolves was perceived as no less than a threat to the rural way of life. Sheep farmers claimed their livelihood was in jeopardy. Potentially declining moose stocks were seen as a threat to the availability of good moose hunting and to the landowners' revenue from hunting leases. Shortly after the wolves appeared, several hunting dogs were attacked and some were killed. And of course there was the age-old fear of wolves. Consequently, conflicts soon flourished. Frank and his friends saw this development not as a mere nuisance that affected their hunting but as part of a development that threatened the totality of their lives. Since they badly want to change this situation, Frank's network is united around a goal that is clearly political: to remove the wolves from their hunting grounds. This would require not only significant changes in Norwegian environmental policy and legislation but also that Norway demand exceptions from international treaties like the Bern Convention. There has been little progress in this direction, to say the least.

In order to explain why our young men are unable to influence a political issue that strongly concerns them, we draw heavily on Paul Willis's seminal book *Learning to Labour* (1977). Willis's object of research, reproduction of class relations through the school system, was rather different from ours. Nevertheless, his informants, "the lads," resemble our hunters in many ways: they were working-class youngsters who unintentionally determined their life trajectories through their own actions as agents of an oppositional counterculture. Although widely regarded a classic, Willis's study from the troubled industrial city "Hammertown" in the 1970s has been the subject of debate since it was published, and many scholars have been critical of Willis's analysis. This critique has focused on issues such as the element of resistance in the lads' behavior, the generalizability of Willis's findings, and the study's relevance in a "postindustrial" society (for an overview, see Arnot 2004). We will not engage in that debate here but rather concentrate on elements in Willis's work that we have found extremely relevant in our studies of the rural working class, namely the concept of "cultural resistance" and its counterpart, "the Hammertown mechanism," a term we have chosen in order to denote the marginalization that may result from "victorious" cultural resistance.

In fact, we will argue that the theoretical scope of the Hammertown mechanism is broader than reflected in the bulk of the literature. As far as we can see, this particular part of Willis's work has been tied almost exclusively to schooling and the reproduction of class relations across generations (see Dolby and Dimitriadis 2004). However, we posit that this mechanism is of a very general nature: successful cultural resistance *generally* tends to perpetuate domination. Drawing on our data from rural Norway, we will try to demonstrate why. Therefore, we think our seemingly exotic wolf example can also contribute to the broader theoretical discussion.

CULTURAL RESISTANCE

The term "resistance" is frequently used in ethnographic studies of the working class. However, quite a few contributions rely on a rather intuitive understanding of the term, especially when the "resistance" is of a somewhat subdued kind. For example, there might be mention of "underlying elements of resistance" or "undercurrents of resistance" (e.g., Evans 2006; Lareau 2003), which is not necessarily a problem in texts that do not have resistance as a main focus. However, as Sherry Ortner (1995) notes, studies that do focus on resistance are frequently prone to a certain simplification of the element of opposition in the everyday practices of "subalterns." There has been a tendency to depict "resistant" action as more coherent and directional than is justified, since people's practices are normally complex and marked by ambivalence and uncertainty (Ortner 1995). In line with Ortner, we see a need to situate elements of resistance in the complicated web of everyday life, that is, accomplish a "thick description" of resistance in all its diversity while also attempting to establish a criterion for what the term can meaningfully cover.

To interpret all cultural expressions different from a dominant culture as resistance is not reasonable. One criterion could be that the resistance, at one level or another, must be intentional: resistance against some form of power must be part of the meaning that individuals attribute to their own actions (Fegan 1986). This means most cultural expressions *can* imply resistance, if only people see their practices and values as oppositional in the sense that they contain elements of conscious defiance against groups that claim superior knowledge and legitimate taste. The concept of cultural resistance takes as its point of departure a relation of power, and it denotes a situation where those in a subordinate position make use of cultural means to challenge domination. Even if these conflicts are most visible at a cultural level, there is an underlying material basis in an uneven distribution of economic resources and power. Concrete cultural resistance springs from a social hierarchy and thus entails a link between social positions and cultural forms. Cultural entities, symbols and signs, values and meanings are all socially embedded, and they vary among hierarchically ordered social positions.

Hegemonic cultural forms and a hegemonic "world view" are met with various counter-interpretations that thrive in the background but are also, to varying degrees, taken out into the open. James C. Scott (1990) writes that subordinate groups create hidden discourses that represent a critique of power spoken behind the backs of the dominant. He terms these discourses "hidden transcripts." While generally "hidden" from the powerful, they comprise interpretations that explicitly defy hegemonic discourses. Cultural resistance is not necessarily launched against institutionalized power and does not generally imply a desire for fundamental social change, but it should be seen as a struggle for autonomy—as an attempt to clear a space out of the reach of power, where one is the master of one's own life. This (potential) autonomy does not in itself entail any corresponding influence outside the cultural realm. Indeed, the opposite may be more likely.

WORKING-CLASS YOUNGSTERS IN CULTURAL REBELLION

During the 1970s and '80s, authors like Dick Hebdige, Phil Cohen, Stuart Hall, and Paul Willis published several works that presented a perspective resembling the one we have suggested here. They described how different subcultures sustained core working-class values but through styles and actions that provoked *all* levels in the class society (cf. Hebdige 1979). Subcultural rebellion could affect the life courses of young people, but it did certainly not affect the distribution of wealth and power in Britain. The structures of capitalist society remained unaffected by the cultural resistance of British working-class youth. Indeed, this form of resistance contributed to the reproduction of class relations.

Likewise, more recent contributions have described oppositional subcultures that succeed in creating a "parallel universe" with its own codes and norms but also contribute to the perpetuated marginalization of their members. Philippe Bourgois's (2003) study of Puerto Rican drug dealers in East Harlem is an often-cited and convincing example. However, in our view, Willis's work from 1977 represents a systematic approach to the mechanism of marginalization through "successful" resistance that is less developed in other contributions. Therefore, we will take Willis's model of the Hammertown mechanism, as we understand it, as our point of departure.

"HAMMERTOWN"

The following is probably one of the most widely used quotations in modern social science: "The difficult thing to explain about how middle-class kids get middle class jobs is why others let them. The difficult thing to explain about how working-class kids get working class jobs is why they let themselves" (Willis 1977: 1). The quotation opens the now classic book *Learning to Labour; How Working Class Kids Get Working Class Jobs*. Willis addresses one of his main points in these first two sentences: people are not driven into manual labor through open coercion. When social reproduction still sorts working-class kids into working-class jobs, this must have something to do with their own actions. By means of a rich ethnography, Willis reveals his core insight: a social mechanism that explains "why they let themselves." The study is based on a small sample of teenagers—a group of rebellious working-class boys ("the lads")— and a smaller group of conformists—"the ear'oles" (earholes), or pupils who always listen to the teachers. He observed them through their last years of school and as they made the transition to the labor market, and he observed how they created a counter-school culture that gave them a specific form of autonomy.

In terms of troublemaking, the lads belonged to the absolute elite. They obviously did not value academic achievement. Instead, they found meaning in disrupting classes, terrorizing teachers, drinking, stealing, and fighting. In combination with a variety of countercultural elements, including outspoken racism and sexism, they cleared a space for themselves, in sharp contrast to core values of the school system. Consequently, they managed to establish an autonomous sphere where the lads were the rulers and where the teachers were off limits. Hence, compared to the ear'oles, the lads were in a sense powerful. Unlike young people who obey authority and absorb the knowledge and values of the school system, they were masters of their own lives. However, their oppositional actions forced them to face a boomerang effect. An unintended consequence of deliberately choosing to ignore school was that they effectively channeled themselves into the lowest strata of the working class. The lads opposed a school that served the interests of capitalism. In doing so they achieved two things: they created autonomy for themselves in relation to the school system but at the same time perpetuated a fundamental mechanism that maintains the social reproduction of an unequal distribution of power and wealth.

The "Hammertown mechanism" unfolds at three levels:

At the first level we meet young people who assign purpose and meaning to their oppositional actions. The lads are not blind victims of the forces of capitalism. The structural constraints of contemporary societies work in a subtler manner, and the power structures of capitalism do not only act as forces from outside of the individual. Willis shows how people can confirm and reproduce structural constraints through active and deliberate everyday practices. The lads are losers within the school context, due to their own interpretations and their own active actions. At this level Willis's book focuses on individual motivations. The school heralds the message that a good performance will provide great career opportunities, which the lads interpret as a falsehood and a deception. For them it becomes important not to look for "interesting" work. In their eyes it is not possible to gain real freedom, or autonomy, by adopting the perspective of the school and seeking a successful career. Instead, they aim for "generalized labor," challenging one of the school's core values.

At the second level we meet the informal social group, where the individuals find resonance for their practices. The group is a necessary condition for cultural resistance, where peers associate and where deep skepticism to the school and its values is effectively formulated. The group's beliefs become core references for validation of knowledge, behavior, values, and moral. Within a network of friends, the oppositional and aggressive patterns of action develop. And indeed, social sciences are familiar with the small informal social group as the place where cultures of resistance are produced. Such groups are described in the works of Sverre Lysgaard ([1961] 1985), Marianne Gullestad (1984), Lois Weis (1990), Thomas Dunk (1991), and Philippe Bourgois (2003).

The informal group operates within the broad context of class relations and unequal distribution of wealth and power. At this third level the oppositional actions find their deepest meaning. Even if the motivational horizon for the lads' actions is local, it basically represents the general relations of domination and subordination that saturate capitalist society. Willis's theory will not hold water unless the lads themselves, on one level or another, recognize these power relations. But that is exactly what they do. The key term is class culture—a class culture that contains many, albeit generally diffuse, insights in societal power structures and in the class relation itself. Willis uses the term "penetrations." Through growing up in working-class families, the lads' thoughts and perspectives become heavily influenced by working-class culture, one that draws heavily on their parents' experiences with manual labor. And even if these insights must be quite vague to youngsters, they are distinct enough to evoke a strong sense of having exposed the hegemonic and repressive ideology of the school as an instrument for domination. Class culture helps the lads to see through the ideological smokescreen of the school system. Furthermore, the values and morals to which the lads subscribe and which motivate and legitimize their actions are variations and recontexualized expressions of the culture that the boys absorbed at home. In this respect, their fathers are particularly important, both as participants in and narrators of class culture and as objects of identification.

This is how power relations in school become variations of the power structures in capitalism, and the relationship between the lads and the school come to resemble general relations of domination. The boys face a patronizing, topdown attitude from the teachers, which is reinforced by their own rebellious actions. Their fathers face the same attitude from the factory management. There is a direct line from the "shop floor culture" (Willis 1979) to the boys' behavior in school. The adult workers think of themselves as the real experts on how to maintain production, possessing the most relevant knowledge and the real power on the shop floor, where everything would collapse if the engineers and managers had their way. Both fathers and sons cultivate strong masculine comradeship. But while the oppositional strategies might function adequately for the fathers on the shop floor, they have negative consequences for their sons in the long run. By participating in the informal anti-school group, the lads obtain not only autonomy but also marginalization. The concept of the Hammertown mechanism suggests that a struggle for freedom within a context of domination and subordination may be successful at one level (in creating autonomy) but at another may lead to strong social reproduction (through the exclusion that the autonomy entails). As already indicated, we see this mechanism as operating at an even more general level, in that it may lead to political marginalization.

We hypothesize that this mechanism also operates in Stor-Elvdal today and can help us understand why the young hunters seem to be sidetracked and powerless in relation to the processes that shape the policies in a field vital to them, namely large carnivore management. We also believe that studying a social group where these mechanisms are easy to identify enables us to understand a phenomenon of a more general nature that normally exists in more subdued forms.

RE-CONTEXTUALIZED WORKING-CLASS CULTURE IN THE FOREST

Like many other rural communities in areas with large carnivores, Stor-Elvdal has regularly been depicted (in the media and in public debate) as a unified stronghold of anti-carnivore sentiments. However, our interviews quickly revealed not only a great variety of opinions on the wolf situation but that those who drew predominantly negative conclusions about the wolf presence also held nuanced and complex views. Furthermore, we soon got the impression that some interviewees adjusted their statements in their dialogue with us, researchers from Oslo with digital recorders. To probe deeper than the normal interview allows, we decided to embark on a limited ethnographic field study, targeting a network of young men who were dedicated hunters and held strong opinions on wolves and wolf management.

Our door opener was Frank, who generously took us home to meet his family and introduced us to his and his friends' everyday practices as young outdoorsmen. During a period of three years, we took part in a series of different hunting and outdoor activities. The men in his network regularly dropped by Frank's house, and we made a habit of visiting him as often as possible. Spending evenings with him, we met his friends and participated in their conversations at his kitchen table. We spent time with them by their campfire in the forest, taking a break from the hunting. Between visits, we stayed in touch by phone, and Frank reported on the general state of things at his end. The young men did not form a social group in any strong sense. No distinct boundaries existed between them and the rest of the community. However, using snowballing as our sampling method, we always asked informants to supply us with new contacts, which usually resulted in the repetition of a limited number of names. This confirmed our impression of a tightly knit network, which was nevertheless fully integrated in the larger community.

The participants were in the process of choosing their life track, and they all knew each other and appreciated each other's friendship. In addition, they were all dedicated hunters, and none of them owned enough land to have private hunting grounds. They all had to buy hunting permits or rely on the goodwill of landowners. Most had fathers who were formerly employed in the local timber industry or as craftsmen. Manual jobs are now far less available than they were when their fathers entered the workforce, yet the men had jobs that did not require an education above the mandatory nine years. Most of them were employed in the public or private service sector. Several, including Frank, had found work in social services or health care. However, some held traditional men's jobs, for instance, as a taxi driver or truck driver.[1]

Hunting is a typically male activity. The network members had learned their hunting skills from fathers and grandfathers, and they saw their outdoor practices as a continuation of a masculine culture. The visible expression of their identities can easily be portrayed as stereotypical male, rural, working class. In significant ways, they shared the lifestyles and values of their fathers, but unlike the older men, most of them maintained traditional masculine working-class identities without traditional male working-class jobs to support them. One of the core qualities of the young men in Frank's network is to simultaneously be a traditional rural man and a modern man. Lars, for example, who works as a truck driver for a timber mill, served us homemade cookies when we interviewed him. He proudly announced that he had baked the cookies himself, using a traditional recipe from the area. Baking, of course, is a traditionally female activity, and Lars's cookie baking suggests he was not afraid to disregard the expectations that young men of his type have traditionally met.

Other leisure activities were often ingrained in local tradition as well. Some of the men made knives, one made traditional wooden furniture, and one was a competent folk musician. They met to hunt and sometimes to spend the night in hunting cabins. On such trips, we have been served beer, homemade spirits, half-fermented trout (typical of the Norwegian inland), salted pork belly, and fatty sausage. When not together in the woods, they met elsewhere, or they talked on their mobile phones about hunting, weapons, dogs, and wolves. The friendship and sense of community that the hunting provided seemed to be important to every member of the network. They lived this significant part of their lives in accordance with what they perceived to be the traditional ways of men in the area.

Nevertheless, those who had started families were modern fathers who assumed considerable domestic responsibility. Frank seemed to be aware of the potential contrast between his life as a rough outdoorsman and his life as a family man. He often made jokes about gender and domestic labor, saying things like, "Well, women, you wash the dishes," or, "We are going to watch some TV; serve us coffee," but they were always followed by laughter. In fact, his family life was conducted in accordance with modern standards: he changed diapers, fed the baby, washed dishes, and scrubbed floors. There were definitely limits to his traditionalism. In his home, we were never served the fatty food we ate on the hunting trips. Instead, Frank and his wife served Italian-, Indian- and Chinese-inspired dishes and "Cajun-crossover-fusion" courses, with chili sauces and garlic. "Do you like garlic?" they asked us more than once. "We love it!"

A WOLF IN THE GARDEN

From building tree houses as children to hunting as adults, the young men in our study experienced the woods as the most central arena for recreation. But this relationship was always on people's terms. Nature was never really wild, although it contained some wild animals. Nature was always a safe playground, a place to roam freely, providing pleasant surroundings for the local community. There is no room for large carnivores in such park-like surroundings. When asked if they meant that usefulness to humans is the only valid reason for a species to exist, all of the men stated that every species has a right to live in its natural habitat. Most of them believed wolves should not be an exception, but they did not believe wolves belonged in Stor-Elvdal. Besides, they pointed out, wolves are not an endangered species globally. Wolves simply did not fit their image of nature in their own immediate surroundings. Wolves threatened to break down the whole concept of what nature in Stor-Elvdal is meant to be: a safe playground for people and dogs and a place where wild game has nothing to fear from species other than humans.

The young hunters interpreted the wolves' presence as a threat to their life projects. On another level, they seemed fascinated by the animal itself. When they talked about wolves, they revealed an interest and a level of knowledge that went far beyond simple hate. Frank had often borrowed DVDs about wolves, and he put considerable effort into imitating their howls. The wolf has skills that the hunters value and admire—skills they themselves would like to possess and to observe in their own dogs. The wolves are wild dogs. They are also great hunters. No animal, or human being for that matter, could receive a better testimony. The men's rage was not directed at the wolf itself, which they saw as an animal that merely follows its instincts, but rather at the human agents of wolf protection.

THE CITY AND THE ENEMIES:
TWO SIDES OF THE SAME COIN

The introduction to a textbook used in the mandatory Norwegian hunting course explains the historical relationship between hunter and nature to the novice: "Through hunting, modern man forms an alliance with nature and his past." And further: "The ancient Nordic hunting and trapping culture is still alive in our country" (Gjems and Reimers 1999: 14). Several members of Frank's network kept their course diplomas framed on their living room walls. They spoke of hunting with dogs as an "old culture" they feared would disappear. In Norway, hunting is culturally constructed as a very old tradition and as a way to escape from a complex and stressful modern society (Aagedal and Brotveit 1999).

The young men justified their choice of place to live not only by praising the good life in the small forest community but also by denigrating the city. When asked to describe the city as a place to live, they all came up with horror stories about crime, drugs, and traffic. Large cities are unsafe, chaotic, noisy, and packed with social misery. They also emphasized the negative aspects of the city as a physical structure—big ugly houses, crowded streets, and, most importantly, the absence of nature. Frank and the others saw the qualities of the rural community and its natural surroundings as being diametrically opposed to the chaotic and unpleasant nature of the city and their life as outdoorsmen as a negation of city life. They were not revolutionaries, but it seems appropriate to understand their love of the countryside and their skepticism toward urbanity not as mere preferences but as a critical attitude toward the general develop-ment of modern society. Their love of rural life did not express a longing for a better past. Rather, they saw their life as outdoorsmen as a present possibility. In this sense, they could be seen as opposing urbanity and even modernity itself.

"They don't understand how it is," the hunters said, when asked to talk about the pro-wolf lobby. "You should have brought them here, and then I would show them what it's all about," was also a common saying. As argued in chapter 3, the appearance of wolves was associated with urban life, city people, and an urban concept of nature. Through this construction, the wolf becomes an icon of urbanity. In the young hunters' world, this is the ultimate antagonism to the nature they love. The wolf is not a part of real nature but rather an urban implant. With the reappearance of wolves, modern urban life suddenly caught up with the hunters, which is exactly what they had sought to dissociate them-selves from by living in Stor-Elvdal.

POWERLESSNESS, STIGMATIZATION, AND CLASS RELATIONS

The network members talk about an antagonistic relationship between power-ful circles with an urban basis and a powerless group living in rural areas. The urban-rural dichotomy is experienced as a deep and many-faceted conflict. At the core is an uneven relation of power. The hunters feel their rural "view of life" has no effect whatsoever on political institutions. They are the underdogs in a relationship where the dominant are perceived as having power in almost all areas of life. We present a longish excerpt from an interview with Kjell Vidar (who is working in the private service sector), where he expounds on the power to stigmatize that others have, how distressing this can be, and how it limits the hunters' ability to get their message across:

> Kjell Vidar: We are often looked upon as a group apart. We really are. Not that *I* see us that way, but people who have these views that I have, we are seen as a different kind of people, quite simply. You can see that from the way they treat us on TV when we try to say what we mean. We *are* seen as a strange breed.

> Interviewer: Now that I have become acquainted with a number of people up here, I think the picture that is presented in [a national tabloid] and other city papers is quite far from the mark, really.

> Kjell Vidar: That is so true. And most of [the journalists] have never been up here. They have never talked to us. But it is obvious that they are only in-terested in writing about the most extreme people in our community. And we do have some extreme individuals who are willing to break the law to get rid of the wolves and that kind of thing. And that is something they like to write about. But that is only a tiny, tiny group. And it's a tragic situation now.

The daughter of a workmate goes to [the university], and she doesn't dare to admit that she comes from Koppang. She really can't do that. (...) She was really shocked when these newspaper pieces about Koppang [and a wolf culling] appeared. When it finally came out that she was from Koppang, she was harassed [by the other students]. And now she was relieved because she was going to do fieldwork and could escape from the university for a period. It's a tragedy. Because it's clear that most of the students are people from urban areas. And it can tell you a good deal about their attitudes toward us. (...) It's the media that have painted this picture of us, which makes them develop these attitudes.

Interviewer: But it has got to be this recent conflict that has emerged that lies behind...—

Kjell Vidar: Yes, absolutely, it's the conflict that has happened now. Because we are some barbaric morons up here who take our rifles to bed and such things.

Interviewer: Do you feel bitter about it?

Kjell Vidar: Yes, very bitter. Because we have no chance to come forward with what we mean and what we stand for. Because nobody will listen to us; it's not interesting. Because we aren't as bad as they make us out. But then we aren't interesting to talk about anymore.

The hunters frequently use the terms "we" and "them." They allude to some vague others, usually some kind of enemy not identified in specific terms. At a general level we can say "the others" are perceived as different in a broad cultural sense, but they are also the ones who have power. In the excerpt above, the terms "they" and "them" are used primarily with reference to people in the media, but this may be more ambiguous than it seems. Kjell Vidar says, "We are often looked upon as a group apart," but *who* sees them that way remains implicit. The point is, however, that the media have the power to define them and that the hunters themselves have no means to break down this image. That the media picture is biased or simply wrong is irrelevant; their picture seems to inform the public opinion. Kjell Vidar's despair is rooted in a strong sense of powerlessness. To get a message across is impossible for him. If he does not want to present himself as a barbaric moron, then they do not want him. The story about the girl harassed at the university serves the purpose of illustrating how the stigmatization works. That this happens at a university is hardly coincidental. "The university" is a strong symbol not only of abstract and useless knowledge but also of the expansion of urban culture. We can discern a line here, from the national media to the university. The people there have the same patronizing view of rural people. It becomes clearer who the others are: a highly educated urban middle class.

But the hunters don't just sit there and take it. They hit back and do not hesitate to make definitions themselves. The hunters describe wolf lovers as ignorant city people, out of touch with the real world, who do not know how nature works; without any insight in the harm caused by carnivores and totally lacking an understanding of rural life and the meaning of hunting. Here is what Erik thinks typical wolf lovers look like and what really motivates them:

> Interviewer: Do you have any idea about what a typical wolf proponent looks like?
>
> Erik: Yes, they usually have rather long hair and a beard. And they often actually wear lilac scarves as well [lilac scarves being an icon of 1970s radicalism in Norway].
>
> Interviewer: Yes?
>
> Erik: Drives an old car. And are very much engaged in social and political issues, in a way. But—
>
> Interviewer: Do you mean that they are active in left-wing politics?
>
> Erik: Yes, sort of. It is difficult to describe what I mean, but—
>
> Interviewer: Well, part of my reason for asking is that—like you said yourself—there is this very common view that all people who are against wolves are backward "peasants," you know—
>
> Erik: Yes, well, I generally see them as organization activists, who are, well, not exactly [anarchists; mentioning a nationally well-known group of activists from Oslo], but there are many who join organizations just to take part in demonstrations, whatever the issue—
>
> Interviewer: Yes?
>
> Erik: You've got Greenpeace and all that, they get kids to join, and the people at the top don't give a shit, they just think about getting rich.

Erik describes a style, a cultural expression that must be interpreted as the exact opposite of their own "hunter's look." Erik also evidently thinks the typical wolf lover has his focus somewhere else: he is mostly interested in demonstrations. And behind him are people who only think about money. Thus, the hunters also create a stigmatizing image of another group. Their cultural antipode, intellectual city people, is often ridiculed. But there is a significant difference: the hunters are underdogs. The stigma they construct is not effective.

A SENSE OF CRISIS: CULTURAL RESISTANCE

We have seen that the young men interpret the wolves' presence as a serious threat to the totality of their life projects. The resistance they launch therefore involves a sense of crisis at an individual, even a deeply personal, level. What they want to be—their identity—is under attack (see Krange and Skogen 2007b). Their problems are hardly caused by the return of the wolves alone. We have seen that the area is plagued by depopulation and a dramatic drop in employment in resource industries. The wolves cannot be the most important factor making it difficult for our boys to live the life they have chosen. But then their resistance against the animal is closely tied to a more general skepticism. Their love for nature and the rural landscape and their rugged style and traditionalism should be seen as opposing urbanity and even modernity itself. Indeed, the totality of their lifestyle conveys a form of resistance, but in an understated, quiet, and mild form that rarely confronts power head on. Even so, it has significant social consequences, as we shall see.

"THE LADS" AND THE HUNTERS: SIMILARITIES

Several similarities exist between the hunters and "the lads." Both are boys or young men from the working class. Against a shifting background, where many core elements in the class culture apparently have a weakened basis, they still reproduce working-class culture. Even if the labor market in Koppang is in transition, as it was in Hammertown, and grown men thus may seem less useful as role models, we can see that heritage from fathers is crucial. The fathers' culture is founded in their subordinate position within the production system. Even if younger generations live their lives out in a different context in both places, the young men in many ways adopt their fathers' traditions. But this is not a "blind" inheritance. Rather, they adopt a perspective—a motivational horizon—that is one of the preconditions for their own development of a resistance culture. Historically, the Stor-Elvdal community has been shaped by the logging and forest industry, as well as a large working class that derived its income from a physical transformation of the forest and timber resources. This material basis for the local working-class culture and the relationship to nature that followed from it, live on in the young hunters' knowledge of nature and attitudes toward it. But the hunters develop a version of their fathers' culture that does not entail opposition against the big forest owners or the bosses at the sawmill. Instead, their rage is directed against a general development that they identify with the city and the dominance of the educated middle class.

Like the lads, the hunters do not choose occupation according to interests. They do not buy the "self-realization-through-work" idea. That is exactly the type of modern nonsense they denounce. Career considerations do not guide the choices they make. Work is a necessary evil. They opt for "generalized labor." The point is to find a livelihood in the region so they can go on hunting. And in this arena, outside of their working life, they find a collective foundation for their identity—their self-perception as stubborn and free. For both the hunters and the lads, the small group constitutes the framework for meaningful oppositional action. The possibility of autonomy and freedom lies in the informal relationships the young hunters enjoy. Willis's lads find this among like-minded schoolmates, and the hunters find such companionship in their free time, as hunters in the woods. None of them pay attention to the school's propaganda about working hard to get an interesting job later in life. Such a strategy would mean that they subordinate themselves to a strong authority and therefore lose autonomy. Like the lads, the hunters end up with unskilled and poorly paid jobs, but denouncing hegemonic understanding of careers helps them achieve the freedom they long for.

CONTRADICTORY CULTURES AND COMPETING FORMS OF KNOWLEDGE

Stor-Elvdal is marked by considerable class differences. A small group of large landowners has derived great wealth from their properties. At the same time, they control the resource that provides the hunters with a meaningful life: nature itself and the game it holds. Nevertheless, the young men did not conceive this as an important class divide. They described the economic upper class with words they mostly use for people like themselves, like "down-to-earth" and "buddy." According to the hunters, another axis of differentiation is more important. In the following excerpt, Kjell Vidar talks first of a large landowner and then about the group he really feels a great distance toward:

> Kjell Vidar: He has a really huge property. But I just sit down by his kitchen table, drink coffee, smoke cigarettes, and talk about my hunting license for the next season. So there are no class differences to speak of. We are all the same I think. Well, there might be some small class differences, but the worst class divide is between people like me and the academics that are newcomers in our community. If you know what I mean?
>
> Interviewer: Yes?
>
> Kjell Vidar: Yes, I don't think that they necessarily are conscious about it, that they want to keep a distance; it's not that, but there is something about them. They are in the habit of keeping a distance to people like us. So, the class differences in our community go between people like me and the highly educated people.

In this popular sense, class is no longer an economic category. Instead, it denotes cultural differences closely related to education, which is why it is important to our understanding of the wolf conflict. Kjell Vidar's version of the concept of class denotes a situation where people with academic education stand against everybody else. And he has a point. People with academic education do have influence, not least when it comes to policy making and management of large carnivores. Science and technological advances are tightly integrated in the development of capitalism. According to Bill Martin (1998), this has laid a basis for the tremendous expansion of the middle class throughout the era of industrial capitalism and has contributed to establishing scientific insights as dominant in relation to everyday, practical knowledge. Scientists are a group the hunters look upon with great skepticism:

> Frank: No, I think that these researchers [wildlife biologists] talk more or less rubbish—that's what I think. That's my own impression at least. They say that there are so many wolves, but they only sit and push their computer keys and look at some maps, and they forget to take a trip outside to see how many there really can be. At least I think so, but then again I'm sure they know a lot too; I don't mean to say anything else, but I don't think they have a basis for all their claims, I really don't.

There are other people the hunters trust much more:

> Erik: Actually, I have most confidence in people from my own group, hunters and the like, observations that local people make. Obviously there will always be some who tell tall tales and exaggerate; we know them up here. We know who are trustworthy and who tends to—
>
> Interviewer: —brag a bit?
>
> Erik: Brag a bit. Exactly. So, in sum, you can always work out a conclusion. You trust your own observations, you know. We spend a lot of time in the forest and see many tracks, tracks from predators and other game; we can see how the game move at all times.

There is little doubt as to what type of knowledge has the most authority when insights established through the network members' direct interaction with nature come in conflict with "scientific truth." And they doubt scientists reach conclusions only through the scientific method:

> Erik: I have never heard of a wildlife biologist who is against predators. So naturally you think of them as champions of the wolves and bears. It's their profession, so I can understand that they need to protect their livelihood. If there aren't any wolves here they are out of a job.

In addition to the more general cultural antagonisms, this type of suspicion leads the hunters to think that biologists are firmly situated in the center of the pro-carnivore lobby. Scientists are not neutral observers of the wolf population's development. Their interests seem to benefit from having wolves in the forest. Seen in this light, it is no wonder Kjell Vidar feels that the in-migrant academics represent the greatest contrast to him and his friends. In his experience, the wolf lovers belong on the other side of this tangible class divide. Two distinctly different cultures are pitted against each other in the carnivore conflict, and confidence in different forms of knowledge is a crucial element. Due to the dominant position of scientific knowledge, a hierarchical relationship between the two cultures is established. But seen from a slightly different angle, the relationship appears to be more equal. Mistrust in the knowledge of the opponent is just as strong in both camps. The hunters will not let "the powerful" dictate their views on the enemy and the cause.

CULTURAL AUTONOMY AND
CELEBRATION OF THE INFORMAL

We followed Frank for a long time, and his behavior was thoroughly informal everywhere. He treated everyone as buddies—in the shop, in the post office, or on a landowner's front lawn. Even talking to strangers on the phone was never "strictly business" for Frank. He wanted everybody to be acquaintances and demanded that all relationships have an informal level. According to Willis (1979), the working-class culture is fundamentally informal. He saw the oppositional activity of "the lads" as a typical example of antagonism between the formal and the informal. The school is part of a formal structure and exerts a power derived from the state apparatus itself. Its pedagogic principles and sanctions are thus instruments that serve the interests of dominant groups in the general power structure. Cultural resistance, on the other hand, belongs in the informal zone. In the informal social group lies the opportunity to withdraw from hegemony, through practices that defy the dominant cultural forms (Scott 1990).

We have already seen examples of this: rejection of scientific knowledge and confidence in practical lay knowledge follows the same pattern. Lay knowledge about wolves is spread precisely in informal relations and is rarely infected by the truths that apply in the circles that form the dominant carnivore discourse. Traces of the same cultural impulse are in Kjell Vidar's account of how he deals with the hunting permit at the big landowner's kitchen table. And the hunters see formal channels as neither relevant nor accessible when they discuss how to reach their goal: to get rid of the wolves. Lars carefully hinted he had already attempted some wolf hunting:

> Lars: I never walk unarmed in the forest (...) No, and I am the forest a lot.
>
> Interviewer: Yes. I get the picture. Well, maybe it is the only way. I can't say that I see many other options for people like you, I have to admit. (...)
>
> Lars: Frank and me and some other boys have talked about it, that we should try to do something. Letters to the editor or something like that. But we don't know how to do it. We are a big circle of hunters, and not only here in Stor-Elvdal, so I don't think it will be a problem if we really want to do something, and if we have the guts. We can have the whole bunch up here and hunt wolves on our own.

Interviewer: So you think that—?

Lars: I think it can be done. But really doing it is something else. Because personally I am so interested in hunting, and I would really like to be involved in something like that professionally. I want to take courses to become a hunting inspector [in Norway often employed by private landowner associations] so then I can't get involved in anything illegal.

Interviewer: No, if you did—

Lars: No, then I would really ruin my own opportunities!

Several things here need comment. The informal group is the structure where the hunters can find the means to reach their goals. But they would have to act outside of formal channels, through secret and illegal hunting, which would be a felony that could lead to a prison sentence. Still, Lars cannot see any other options within his reach. Frank and Lars had considered entering the formal zone by writing to a newspaper, but they don't feel at home there and lack the resources required of actors in that arena. They simply do not know how to do it.

We see here how culture may improve as well as limit people's influence in areas important to them. Lars describes a concrete example of how symbols and language can affect people's access to political processes. Seen in this perspective, the hunters are victims of a cultural hegemony. We must assume this is not the intention of those who shape Norwegian carnivore management strategies. Attempts at conflict mitigation and public involvement in various processes point in a different direction (Skogen 2003). But the fact remains that the cultural resources or symbolic capital needed to influence formal arenas do not exist in the culture that the hunters master. Kjell Vidar told us about a tangible class divide between him and the in-migrant academics, but he also emphasized that the academics hardly created this divide on purpose. And thus it is still another example of unintended consequences of class-cultural differences. But the example also shows that this is not something that simply happens; that the hunters learn to love their destiny. They are actors who fight back—however inadequate the means—for their culture and their alternative view of large carnivores.

Furthermore, we see that Lars for his part dismisses the possibility of illegal hunting in the end. His ambition is to turn hunting into an even more important element in his own life, which makes him extra vulnerable to the sanctions of the formal power structures. In a sense, Lars and the others have broken loose from the hegemony of the dominant culture. But their cultural resistance is always launched inside a larger context, where the informal resistance efforts are in a subordinate position in the end. This is an important point for Willis as well. The lads don't always get away with breaching the rules. In *Learning to Labour,* for example, when two of the boys steal car radios and get caught, an initially exciting act of toughness and independence turns into something horrible. The meaning originally attached to the thefts, which had developed in the informal group, did not survive the confrontation with strong, formal power structures. In the same way, Lars's joy over having shot a wolf would soon turn into grief if he were caught. This tells us something general about such power relations, because the opposite mechanism is not conceivable. Even if the informal resistance culture penetrates and challenges dominant culture forms, it will never be in a position to break up the hegemony. One reason is because the form of rebellion in which the lads and our hunters are involved does not represent a planned strategy for change or a coherent political alternative. Erik expressed it like this: "We can't do anything about it really. The authorities have the power to decide things. But I am quite certain that quite a few predators are shot when people are out hunting for legal game." He may be right in that some animals are shot illegally, and he is most certainly right in that this doesn't help them solve their problem. Because the authorities *do* make the decisions, and this is an exertion of power that hunters' resistance strategies will never come to grips with. They may shoot some wolves, but that will not in any way change the official goal of securing a viable wolf population.

In the informal zone, a distance to power is created, and such a creation is an important function of the cultivation of the informal. This way, the hunters, like the lads, achieve autonomy. They create a perimeter inside which power cannot reach them but at a cost: the hunters' autonomy—their informal ways of handling everything, the knowledge they rely on, their style and defiant perspectives on life—also prevents them from engaging with the dominant forces in the field of large carnivore management. And this is analogous to the boomerang effect—the marginalization—that the lads experienced. In a sense, the hunters cut themselves off from any influence they might have had on a political issue they consider extremely important: the management of the wolf population in Norway.

"THE LADS" AND THE HUNTERS: DISSIMILARITIES

Our ambition was to explain why the hunters are unable to influence a political process that strongly concerns them. We have answered by referring to Willis (1977); they are trapped by "the Hammertown mechanism." But does the comparison really hold water? Aren't there too many dissimilarities between "the lads" and the hunters? The hunters' situation is indeed different from the lads'. While the lads were schoolboys, the hunters live their lives as young adults in an environment that is by no means as deprived as Hammertown in the 1970s. While the lads were heading for low-paying jobs or even unemployment and general marginalization, Frank and most of his friends keep jobs they are relatively content with and that enable them to live rich lives with spouses, kids, houses, and cars. And they manage to keep up a quite expensive and time-consuming leisure activity, hunting. Hence, compared to the lads, the hunters are not marginalized in a strong sense of the term. The hunters share domestic responsibilities with their spouses, and some have female bosses, which do not seem to be a problem for them. They are not nearly as sexist as the lads. Furthermore, none of them ever spoke of workplace conflicts or frustrations related to power relations at work, which does not seem in line with the lads' experience either. Hunting skills are highly regarded among many of Stor-Elvdal's inhabitants. This provides our boys with a certain kind of respectability and a source of self-esteem and even identity: they *are* hunters. The lads, on the other hand, were to some extent marginalized even in their own community.

Finally, there is of course the question of resistant actions. One significant difference between the lads and the hunters is that the lads more frequently met their enemy face to face. Examples of manifest head-on conflicts with agents of power are few in the hunters' case. However, we do not see such direct confrontation as a prerequisite for actions to be understood as resistance. We have argued that intentionality should be the base requirement. This is in essence a subjectivist understanding of resistance, which, in our view, ties the concept to the production of meaning in a way that has considerable analytical potential. Accordingly, whether an action should be regarded as one of resistance is up to the hunters themselves and the meanings they ascribe to their practices. And, as we have seen, the cultural contrast their lifestyle constitutes vis-à-vis dominant middle-class culture—with its clearly visible ties to political power—in itself has meaning to them: a sense of freedom from domination, that is, a sense of autonomy. This would not have been possible if they did not recognize and appreciate this cultural contrast as something more than diverging preferences. At this level of meaning, the way they perform everyday life is a way of resisting domination. That the "oppressive urban elites" do not really pay attention does not matter. Defending a meaningful and coherent lifeworld and establishing what is deeply sensed as cultural autonomy is clearly interpreted as a form of opposition.

To conclude: the lads and the hunters are similar and dissimilar in many ways. However, similarities on a deeper structural level justify and substantiate the comparison. Let us recall the general pattern of the Hammertown mechanism: within an informal group, actors deliberately develop a counterculture through recontextualization of cultural impulses they have known since childhood. A practice in opposition to a dominant culture is developed. The informal group operates in a larger field of conflict between domination and subordination. In this larger context, a boomerang effect appears: through their oppositional practice the group members place themselves on the sidelines in several respects.

The hunters follow a parallel trajectory: a group of young men puts up resistance against several aspects of contemporary social change. For them the reappearance of wolves has become the most prominent symbol of a development they see as entirely negative. With a basis in local hunting traditions they have known since childhood, they develop a counterculture. In its form and style, their cultural identity represents the opposite of the cultural forms that dominate within the circles of power. They have obtained cultural autonomy. They are "the rulers of their own lives" and do not succumb to cultural hegemony. But this freedom comes at a cost. By defining themselves and their practices explicitly in contrast to the dominant culture, they block their own access to the political processes that determine land use policy and therefore the conditions that frame the lives they want to live. This is the Hammertown mechanism in its most general form: even understated and mild forms of resistance—also at the level of Scott's (1990) hidden transcripts—have social costs. In the hunters' case, they contribute to political marginalization.

DAMNED IF YOU DO, DAMNED IF YOU DON'T

By entering the formal political arenas, if they could enable themselves, the hunters would have to accept and even adopt values and modes of understanding characteristic of dominant social groups. They would then contribute to still another victory of the formal over the informal and lose the precious autonomy they have achieved. Furthermore, a sizeable majority of the Norwegian population want wolves in Norway (Tangeland et al. 2010), and international conventions set the framework for modern large carnivore management. The logical consequence of the hunters' desire to remove the wolves from their own forests must be to eliminate wolves from Norway altogether; otherwise they would only shift the burden to people like themselves in other areas. Such a goal is in reality impossible to accomplish. And since the actors with power in the field of large carnivore management endorse the conventions and would like to see viable carnivore populations, those who seek influence must accept this premise. For the hunters this would mean giving up their goal and losing their autonomy at the same time. And then what would they gain?

THE GENERAL NATURE OF CULTURAL RESISTANCE

This chapter has dealt with a relatively small group of young men who live and hunt in the forests of southeastern Norway, and we think we understand their resistance against wolves quite well. But many more share their political ambition. So could the Hammertown mechanism shed light on opposition to wolves in a broader sense? We think so.

Willis's (and our own) focus was on what one might call oppositional outliers. As a highly visible and in some ways influential group, Willis's "lads" were worthy subjects of study in their own right. At the same time, their spectacular behavior and verbal expressions highlighted processes that are also important in broader social segments. Among the majority, a similar protest might take more moderate forms but can be better understood in the light of analyses of particularly clear expressions. Furthermore, such studies may challenge our habitual notions: the main reason the lads did not do well in school was not their lack of resources. The simple fact that not all do the best they can has been an important insight in education research.

In much the same way the story about the lads was broadly relevant to education research, our story about Frank and his friends resembles other accounts. Since we did our fieldwork in Stor-Elvdal, we have studied the conflict over wolves in many other communities. Everywhere, we heard similar stories and the conflict patterns were the same. The power relations framed within the triangle of resistance, autonomy, and political marginalization occur everywhere. We have seen this in the conflicts over large carnivores, but the same pattern is obviously found in many other conflict areas. Wherever we have presented our findings and suggested interpretations, people nod in recognition—whether in Europe or North America, or even in India. We are certain the stories from Norway could have been recounted from many countries where societal development trends in one way or another lead to more encounters between humans and carnivores and at the same time supply particular frameworks for understanding these encounters. The conflicts are not local in their nature, although they are played out locally. Therefore studying a small group closely can enable us to describe social processes of a quite general nature.

NOTES

This chapter is a revised and extended version of Krange and Skogen 2011.

1. Even the unskilled segment of the Norwegian working class is relatively affluent, compared to many other countries, and the public sector continues to absorb many people who can no longer find jobs in manufacturing, agriculture, and so on. Yet, unskilled workers are usually at the bottom of the wage ladder, even here. But the low cost of housing in declining rural areas gives our boys a "home advantage" they would lose if they moved to an urban area.

CHAPTER FIVE

SOCIAL REPRESENTATIONS
OF THE WOLF

✳ ✳ ✳

We have said that the relationship between lay and scientific knowledge is a key element in conflicts over wolves, so now we will take a closer look at lay understandings of the wolf and how this kind of everyday knowledge plays into the controversies surrounding the animal. The theory of social representations is a theory of knowledge that explains lay understandings and how they can be studied. Based on this theory, we will attempt to extract some core elements of widely held ideas about what sort of animal the wolf is. These are worth investigating not least because decision-makers and journalists often appear to assume that divergent attitudes toward the wolf always go together with different notions of the wolf as an animal. In particular, people who object to the presence of wolves in Norwegian forests are usually understood to have a dim view of the animal itself.

We need to ask what it means to be "for" and "against" the wolf. Does it mean, for example, liking or disliking the animal itself, or its presence in Norway? People are not born with attitudes. Attitudes are simplified categories that only reveal something about somebody's understanding of a situation in a given time and place. Disputes over ideals and values in the real world are usually far more complicated than the opinion polls suggest with their "for" and "against" response alternatives. They cannot be meaningfully described in terms of dichotomies like "negative" or "positive," "agree" or "do not agree." We have already touched on hunters' ambivalent relationship with the wolf. Many hunters are clearly staunch opponents of the wolf, but they still respect and admire the animal. We could generalize this example. The fact that people can have different opinions about the same issue—whether Norway should have wolves, for instance—does not necessitate similar variation in understandings of reality at a deeper level. Therefore, whether supporters and opponents of wolves in Norway also disagree in their perceptions of the animal itself is an open question.

To find out whether the social conflict lines up with different perceptions of the wolf per se, we need to step back from the conflict perspective. In this chapter, we want to take a closer look at people's thoughts about the wolf, not simply whether or not they want the animal in Norway. Our analysis is based on focus group sessions conducted in Trysil and Halden in 2007 and 2008. Group sessions were arranged with farmers, sheep breeders, landowners, hunters, conservationists, hikers, mushers, local tourist operators, teachers, architects, nurses, sawmill workers, and neighborhood groups. None of the Halden informants came from the town itself, and many lived either in or near wolf territory.

We can divide these groups roughly into three basic categories. The first consists of people who dislike the presence of wolves in their area (hunters, farmers), the second of people who want wolves in Norway (conservationists, one group of neighbors), and the third of people with a variety of opinions (hikers, the second group of neighbors, mushers). The purpose of the analysis is to shed light on the associations the wolf evokes among not only supporters and opponents but also those with neutral or ambivalent positions. First, though, we must say something about how we may go about studying conceptions.

SOCIAL REPRESENTATIONS

Investigating people's understanding of wolves, rather than just the opinions they express, requires suitable tools. To this end, our study uses the framework of social representations as developed by the French social psychologist Serge Moscovici. Inspired by his work, social scientists from diverse fields have studied, for example, social representations of illness (Herzlich 1973), the human body (Jodelet 1984b), cities (Milgram 1984), biotechnology (Gaskell and Bauer 1998), and the environment (Félonneau 2003). In line with social constructivist thought, as exemplified by Peter Berger and Thomas Luckmann (see the introduction), Moscovici posits that all social interaction presupposes cultural frames of reference, enabling shared understandings. Unlike many colleagues in his field, however, he describes the contents of these cultural frames of reference in some detail. People's opinions on a specific matter, he contends, are only a tiny fraction of a larger complex that consists of shared, often implicit, notions of the physical and social world in which we live (Moscovici 1969). These representations can be studied as a set of references we use to orient ourselves in our social and physical environment. In this connection, Moscovici often refers to "common sense" (e.g. 1969, 1976, 1993).

A social representation can be described as "the collective elaboration of a social object by the community for the purpose of behaving and communicating" (Moscovici 1963: 251). This means social groups tend to develop their own interpretations of phenomena seen to be important for some reason, for example, because they are perceived as threatening. By a process of communication within the group, a shared understanding of the phenomenon emerges. This understanding gradually becomes part of the group's implicit everyday knowledge. As Moscovici puts it, "the purpose of all representation is to make something unfamiliar, or unfamiliarity itself, familiar" (2001:37). As we shall see, much of the wolf debate rests on a shared assumption about what constitutes a wild animal. Such assumptions are elements of a repository of everyday knowledge so firmly embedded in a shared culture that lay people, scientists, and managers alike take it for granted.

Like many other social constructionists, Moscovici wants us to differentiate between the reality that surrounds us and our understanding of it. The latter must be our focus when exploring people's relations with their surroundings. The theory of social representation therefore represents a critique of the classic psychological approach to lay understandings, an approach that continues to inform research, policy making, and management (whether we are talking about large carnivores, public health, or climate). In the case of the wolf, we are thinking of an apparently widely held idea that people's conceptions of carnivores are either "right" or "wrong," "understood" or "misunderstood." The problem is not that this approach assumes there are right and wrong versions of reality but that we fail to pay attention to versions other than the one that has achieved status as the "correct" (Jovchelovitch 2008). This failure is problematic because sociological studies of people's relations to carnivores are supposed to tell us why the conflicts become so intense. Insofar as social groups usually think and act in accordance with their own conceptions of how the world looks, we obviously need to look at those conceptions, irrespective of whether they are "right" or "wrong" (Bauer and Gaskell 2008).

Moscovici wants us to approach popular knowledge and conceptions as interesting in themselves, not simply as misunderstandings or as more or less correct representations of scientific knowledge. He therefore describes social representations as "cognitive systems with a logic and language of their own ... They do not represent simple 'opinions *about*,' 'images *of*,' or 'attitudes *toward*,' but autonomous 'theories' or 'branches of knowledge,' for the discovery and organization of reality" (1969: 10). In our study of the social representations of the wolf, we wanted to let lay knowledge be expressed on its own terms. Applying the concept of representations also directs our attention to shared knowledge. Insofar as representations find expression in a collective, they need to be understood as social. Representations are social because they are shared by a group of people, regardless of its size. Inasmuch as the boundaries between social groups are flexible, and since one person is usually a part of many groups—smaller groups, larger groups, societies—the degree of consensus about interpretations of the world must also be adaptable. This is why Moscovici and his colleagues describe the social representation of an object or phenomenon as ideas at different levels. They distinguish between the central and peripheral components of representations. Popular knowledge of a phenomenon consists of an assembly of beliefs, evaluations, and attitudes (Abric 1984: 180), each mutually dependent on the other and hierarchically organized. Some are central and basic; others are more peripheral and mutable.

Any representation could be seen as organized around a central core of basic convictions (Wagner and Hayes 2005) that derive from a common set of social conditions, culture, and history, and consist of nonnegotiable ideas. They are stable, perpetual, and consensual. We are talking about a form of culturally rooted conceptions that are self-explanatory and self-evident. Whatever other opinions they might have of the animal, most people would take for granted that the wolf, as a species, is wild. A representation's peripheral components provide the glue that binds the group's interests and concrete experiences to basic convictions in a meaningful way. They are adaptations of the representation's core components to a particular situation. In contrast to the core components, peripheral ideas are flexible, heterogeneous, and able to withstand internal contradictions—both between group members and in the mind of an individual member (Abric 1993). An example of a peripheral component is the idea held by wolf opponents that the wolves living in Norway today are "unnatural" or "fake." We shall show later how this idea creates a logical bridge between the basic conception of the wolf as a wild animal and negative opinions of the wolf's presence in Norwegian nature.

At one level, this is about basic ideas shared by many, which can be difficult to identify in public exchanges for precisely that reason. At another level, thoughts about the same thing or phenomenon are divided. This is where negotiations or conflict take place, between social groups and between individuals who fight for the right to define reality. Are the wolves living in Scandinavia authentic, threatened wild animals with a rightful place in pristine Norwegian forests? Or are they Russian intruders that corrupt Norwegian nature and threaten other (native) animal populations—domestic animals as well as wildlife—and thus also rural livelihoods and quality of life? If Moscovici and his colleagues are correct, to simultaneously agree and disagree about the same issue is entirely possible, as long as the levels are kept apart. This is an important insight when we set about understanding social conflict, and a good starting point for studies of people's opinions about carnivores.

COLLECTIVE REPRESENTATIONS OF THE WOLF

Although the debate about large carnivores leaves us with an impression of stark antagonism between hunters and farmers on one side and conservationists and "wolf enthusiasts" on the other, our studies show that their conceptions of the wolf are fundamentally the same. Based on the characteristics our informants collectively emphasized when talking about wolves, the predator emerges as superior, social, wild, and pure—ideas shared by most of our informants. In fact, the descriptions across focus groups were astoundingly similar. Words like "genuine," "pure," "unpolluted," "smart," "socially intelligent," "strategic," "dominating," "beautiful," and "magnificent" were repeatedly used, both spontaneously and when informants were asked to describe in their own words the essential nature of the wolf.

None of these attributes were subject to discussion or met with objection in any of the groups. Other participants' body language, such as nods and collective expressions of agreement, underscored the traits' "goes without saying" quality. In terms of what we have said about the central elements of a representation (cf. Abric 1984, 1993; Wagner and Hayes 2005), these attributes seem to be core elements of the profile our informants drew of the wolf. In line with Moscovici's conceptual language, they belong to the representative core, which was confirmed by the way some attributes acted as final arguments in the interview situation. They were the underlying descriptive elements that justified and explained other assumptions about the wolf (Wagner and Hayes 2005: 193). During the interviews and subsequent analyses, specific components of the representation's core were clearly activated in discussions about carnivores. These were particularly relevant and self-evident ideas, which in specific situations were presented as obvious reasons to think about the wolf as a wild, pristine, undomesticated animal. In what follows, these attributes will therefore claim most of our attention. The idea of "the wild" in particular helps explain the opposing attitudes toward the predator.

THE SUPERIOR WOLF

"You'd have to search a long time to find a more fascinating animal!" we were told by a hunter from the Halden area and a declared opponent of the wolf. His words typify how the large carnivore impressed many of our informants, who described the wolf as intelligent and strategic, an excellent hunter, an animal that stands out in terms of physical stature. Many informants, including the skeptics, also saw the wolf as an exquisite, majestic creature, towering head and shoulders above most if not all other species in the Norwegian fauna:

A: It's a bit dignified, maybe.

B: I think so too. They like to be in control in a way.

A: And when you see—I've never seen wolves, but I've seen tracks—and when you see how it digs in the ground, it's impressive. Like shoveling the snow away. It's so powerful. Like a force of nature! And it can leap about five meters ... In my view, the wolf is sort of a pinnacle in the natural world. I'm not blown away looking at wolves in a zoo, but when you know wolves have been crossing the [frozen] lake, you can't help grinning to yourself while you're sitting there drinking your coffee, you know.

(Conservationists)[1]

The idea of the wolf's supreme position in the animal kingdom is reflected in conceptions of its superior physical strength and agility. During the focus group sessions in Halden in 2007, many recounted close encounters with wolves. Even more informants from Halden and some from Trysil[2] had seen wolf tracks, which, in their opinion, were an important source of information about the animal. "A wolf track, you could say, you'd talk about it much more than a moose track," one of them explained. By reading tracks in the snow, they imagined scenarios where the wolf could put its agility and strength to the test, such as leaps of five meters or more on the frozen lake in winter or two meters over an upended tree's gigantic roots to catch a roe deer sheltering from the wind on the other side. Other scenarios described the wolf's unparalleled success as a hunter:

Once, we were tracking two wolves (...). And then, apparently, one of them had stopped, standing still on his post. We could see that he had been standing there for quite some time, waiting, while the other one kept on, and then chased a roe deer right up to where the first one was waiting. Then [the first wolf] brought down the roe deer, and they ate every little bit, except the antlers and the skull. That's all that was left. I was so impressed when I saw how they had been working. This was really someone who knew how things should be done!

(Neighbors group 1)

Like the informant who told this story, many were amazed by the predator's cunning and capacity to work together on a hunt. "Intelligent," "smart," and "strategic" were frequently used to describe the nature of the wolf in general and its hunting prowess in particular:

Interviewer: You said in the beginning that the wolf in your opinion is an intelligent animal.

A: Yes.

Interviewer: Do you others agree?

B: Smart, absolutely. [Several mumble in agreement]

C: If you're going to hunt successfully and live on what nature provides, you can't not be smart.

(Farmers group 1)

Informants in various focus groups remarked how the large carnivore took advantage of tracks left by humans, such as dog sledding trails or plank piers in the marshes. Many were amazed by the wolf's ability to place its feet perfectly in the hollows left in the snow by the wolf in front of them to save energy:

> Indeed, we've seen how fine hunters they are—and how rational they are, actually. We don't often have deep snow here, but it sure happens. And then we can see that four of them have been following each other, or five—but it's totally impossible to decide how many there are, because they're treading exactly [in each other's footprints], saving their strength, you know. It's just incredible!
>
> (Farmers group 2)

The wolf seems to enjoy a special position even among the carnivores. Along with being an impressive hunter, it also consumes, apparently, enormous quantities of meat. It needs to be good at hunting, according to our informants, to satisfy its huge appetite:

> A: For pity's sake, how much meat do they put away? I've read it somewhere. It's not trifles!
>
> Interviewer: You are right, it's a lot. I can't remember offhand, but they consume considerable quantities of moose and roe deer.
>
> A: It's huge quantities. In fact it's so much, I've sort of asked myself, "Can it be right?"
>
> (Hikers)

The predators not only consume many individual prey but apparently eat up every last morsel and never leave a half-eaten carcass. Unlike lynx, for example, which, according to some informants, will not touch their prey's stomach; the wolf is far from picky in the culinary department:

> A: And the wolf, it devours the whole roe deer. No leftovers, neither hide nor hair.
>
> B: No, it takes skin and bone—of moose and ...
>
> A: Well, they say nothing goes to waste, neither bone nor shag nor hoofs nor anything.
>
> (Neighbors group 2)

Of note is that this conversation took place when the topic was the predation of wild animals and does not conform to the informants' picture of a wolf "running amok" in a flock of sheep. When wolves encounter domesticated animals whose natural survival instinct is long gone, and who lack even the sense to run, the wolf loses its instinctive control, our informants believed. It turns into a killing machine, a victim of his own instincts. However, the wolf's appetite for flesh attracted neither condemnation nor admiration; it was depicted as a natural urge or life force. No one, in other words, suggested the wolf was particularly malicious or rapacious because it helps itself to the available resources. Whether the wolf is seen as a threat to whole populations of other wild animals completely depends on how robust the populations are assumed to be.

THE SOCIAL WOLF

Given these conceptions of the wolf's intelligence, strength, and hunting skills, many of our informants unsurprisingly dwelt on the animal's powerful survival instinct. As a species the wolf was considered extremely tough and capable of surviving in inhospitable environments. The ideas about the wolf's survival skills, superior physical strength, and capacity for strategic behavior appeared to rest on a perception of the wolf as a social animal:

> Interviewer 1: If you were to describe the wolf, how would you characterize that animal?
>
> A: Social.
>
> B: Intelligent.
>
> Interviewer 2: How ... can you expand on that?
>
> B: Intelligent? Well, at least it has a social intelligence. It's a very social animal, and it's capable of making—well, strategies while hunting.
>
> (Conservationists)

The wolf's astonishing appetite for meat was intimately connected to the idea of the resilient, energetic pack. Many wolves will need a lot of food, and the predator's combined hunting skills and collective spirit gives the pack what it needs to provide enough food for its survival. And, as we have already seen, the wolf is no mean eater:

> Apart from that, we've had lynx here, passing through They don't take much, and it's something you probably have to put up with anyway. Wolf packs, that's another matter. I think it would change the climate [of opinion] altogether. If one or five roe deer disappear—we've a sizeable population round here—it doesn't make much difference. If that was all it was, I'd go along with the dialogue about diversity, see? But say a pack of wolves turned up in the woods hereabouts—they'd clean the place out before you knew it!
>
> (Farmers group 2)

This, it seems, is how things hang together: the attributes of the pack make the wolf invincible. "What is it about the wolf," one of the hikers wondered. "What is it about this carnivore that gets people so worked up?" Others asked similar questions. According to our interviewees' descriptions of the wolf, the animal's social qualities and loyalty to the group make it so special—an animal out of the ordinary—and set it apart from other large carnivores:

> Interviewer: You said something about the wolf being a special animal, or quite unique. Do you remember—a little while ago?
>
> A: Yes, it is special—the way it behaves. It's a very social animal, you know. The lynx, for instance, is more of a loner, or a part of a family, as long as there is a family. Then they separate gradually, I guess. The bear, too, is mostly living by itself.
>
> (Farmers group 2)

Much was said in all of our focus groups about "packs," "groups," and "territories," and the interviewees' choice of words indicates a conception of the wolf pack as a sort of autonomous, cohesive organism. The potential of the components—that is, the individual animals—derives above all from membership in the larger unit. The essence of the wolf is articulated through the group. As a team they show themselves to be the efficient, strategic, and determined animals they are. In other words, the group is understood as the superior wolf's "normal state." In line with this, our informants see solitary animals, or stray wolves, as individuals dislocated from their natural existence as part of a group. As one of the mushers remarked, "Lone wolves maybe look a bit weird because the wolf is a very social animal. If they're alone, they can start behaving in odd ways. But if there's a pack, a safe and stable group, I think it's a more normal situation for them."

By virtue of being one among many, a part of a streamlined machine, the wolf seems almost impossible to eradicate. The territory or home range[3] is understood as more than a space for a gathering of individuals: it is the self-perpetuating fountainhead and vital force of the species—where they breed, where missing elements are immediately replaced to maintain the group's internal balance and outward power. If one is removed, a new one turns up without delay:

> Think of an established alpha pair with cubs, and perhaps a pack of five to ten animals, it's meaningless ... It doesn't make sense to start thinking, "damn it, I just got to get rid of that wolf." You [remember] the one that was run over and killed, or the one that was shot ... and a year later there's a new one, and a new male arrives too, and ... new ones arrive and take over the territory, and as long as the territory is still there ... You don't get anywhere by shooting a wolf in an area like that.
>
> (Farmers group 2)

Not only do the wolf's social characteristics help maintain stable packs; our informants also saw the pack as an extremely efficient reproductive unit—yet another manifestation of the wolf's capacity to adapt and survive. Ideas about rapid reproduction rates and dispersal were firmly entrenched among our informants. The carnivore's vitality and potency are understood as explosive and potentially uncontrollable, in a human perspective—like an undetonated bomb. A hiker compared the wolf's reproductive capacity to that of mice:

> I read something not long ago ... It was probably in the paper, or maybe on the TV, I'm not sure. Whatever, they'd estimated that this year, the Norwegian-Swedish wolf population had produced about a hundred cubs. If they're right, and we don't do anything about it, in two to three years—how many cubs could there be then? It could be thousands, in no time. It'd be an explosion, you know, if nothing's done. It'll be almost as bad as it is with the mice.
>
> (Hikers)

What is more, some of our interviewees asserted that the internal hierarchy of the wolf pack unavoidably leads to geographic dispersal and establishment of more and more new territories. In a family group consisting of an alpha female and an alpha male, there is no room for other leaders, and only these two get to mate. The strict distribution of power in the pack becomes instrumental in the species' dispersal, since adult individuals must leave the pack to form new families:

> A: But a group like that, it can't just grow and grow forever. Isn't it the case that when it gets to a certain number of individuals, they split off, or at least leave and create new areas—isn't that how it works?
>
> B: Well, they probably wander off when they're mature enough.
>
> (Farmers group 2)

In addition to most distinctly setting the wolf apart from other large carnivores, the idea of the pack or group clearly creates a more or less evident sense of anxiety among the informants:

> Interviewer: But do you think it's dangerous and scary—what do you think—for people?
>
> A: Yes, it sure could be if there's no other food around.
>
> B: And if they're in a group, I'm thinking—if there's a crowd of them.
>
> A: You're right there. Yes, there's no doubt about it.
>
> (Neighbors group 2)

Any fear of the carnivore that our interviewees expressed connected to the idea of the rapidly growing and invincible pack. The sound of a lone wolf howling at the moon, a fleeting glimpse of an animal rushing into the bush, or traces in the snow from a wandering pair—these evoked curiosity and were seen as harmless events. But the idea of multitudes of wolves could be associated with something creepy:

A: To look at, it's a magnificent beast.

B: A little scary.

C: Yeah, a little scary and a little unpredictable, and maybe a little dangerous too if they come several at a time.

(Neighbors group 2)

Both wolf supports and opponents in all of our focus groups expressed this skepticism. Even the conservationists, who would like to see a larger wolf population in Norway, felt uneasy about the prospect of uncontrolled population growth:

The fact that the wolf creates anxiety ... like we discussed a minute ago, and ... that's a fact ... they're probably afraid it will multiply, maybe much more than it does now. I don't know how fast it breeds, but it goes without saying, if there are huge numbers, there has to be some sort of control—that's my opinion, anyway.

(Conservationists)

Contrary to the impression the nationwide carnivore debate might have created—that some people hate wolves and would prefer the whole species be eliminated from the face of the earth while others cannot get enough of them—people in this study agree that breeding and dispersal must be controlled. However, they are divided on how many wolves the country should tolerate, or even encourage, and where they should be allowed to settle. People's unease about unconstrained breeding needs to be understood against the background of the core attributes of the social representation of wolves that were discussed above. The animal's preeminent strength, intelligence, and numbers evoke the potentially frightening idea of an uncontrollable population, beyond the reach of human interference.

THE WILD WOLF

Of the typical attributes our informants highlighted, the picture of the "wild" and "pure" wolf stand out. Given what we have already said about the core elements of a social representation, these traits appear to occupy the center of the wolf's social representation. The notion of the wild was expressed in several ways, generally in contrast to the "socialized," that is, whatever is affected by human control and influence:

A: I know they experimented using wolves to pull sleds in Alaska many years ago. And there was one there who had raised several cubs, which he intended to hitch up with other dogs. But it didn't work because they couldn't control their diurnal rhythm. So when it grew dark, the [wild] wolves wanted to push ahead all night at full speed. It was during the day *he* wanted to travel. They just lounged around sleeping. So the diurnal rhythm was out of sync, and he couldn't fix it.

B: That's true. Quite a few people have tried it in Alaska, but none successfully.

A: Well, I know they could have with a mixed breed.

B: Yes, the more dog it became ... a quarter breed.

C: A quarter or an eighth.

B: They gave up in the end because it wasn't worth the effort.

(Mushers)

In this instance, the domesticated dog represents the socialized, contrasting and underscoring the wild nature of the wolf. Our interviewees point to one element of the wolf's nature, its diurnal rhythm, to exemplify what they believe is the wolf's immunity to human influence and control. The wolf's instincts, or innate characteristics, are apparently so resistant to domestication that a large percentage of domestic dog must be in the mix before it can be involved in any human activity. The wolf's wildness is seen as such a strong aspect of its nature that the animal just cannot be tamed.

Others, however, see the wolf as more capable of adapting to humans, as illustrated by accounts of wolves prowling farmyards and lurking outside day care centers. In any case, socialization of the wolf would result in the loss of its wildness. A tame wolf stops being a "real" wolf. When someone in a focus group jokingly suggested it would be better to build fences around the wolf rather than the livestock, a representative of the local farmers' trade organization responded: "At the very least, you'd be dealing with a changed wolf, more of a dog. A wolf must hunt, one way or the other. A wolf that's fed isn't the same wolf anymore." While the choice of words and examples may differ, everyone agreed a real wolf is essentially wild.

So how is this wild quality expressed? On its own, the idea of "wild" is an empty category that acquires meaning only when juxtaposed against its mental opposite: the socialized and the human. The hallmarks of the wolf as a predator are precisely its autonomy and independence of humans. Its qualities as a hunter only affirm the wolf's autonomy. They give it the means to feed itself and its young, and they attach the wolf to areas that ideally lie beyond our reach, where it can avoid conflict with the world of humans. This is how the predator is perceived in its wildness, both in essence and because its surroundings, the "wilderness," are non-socialized. An autonomous animal that belongs in non-socialized surroundings would be expected, reasonably enough, to be cautious in dealings with humans and their physical domain:

Most people, from what I gather, may have felt it was exotic and exciting to go and look for tracks and ... even getting a glimpse of them just that once, but it's extraordinarily difficult. They're so shy, and they never come close to the houses round here.

(Neighbors group 1)

Sightings of wolves near residential areas are described as unexpected forms of boundary crossing. In contrast to the wild, people's homes represent the geographical and symbolic core of the socialized sphere. Describing an encounter with a wolf on a forest road, a conservationist put it like this: "I thought it was a dog; I hadn't expected to meet a wolf three hundred yards from my front door!"

As mentioned, some of our Halden informants—particularly those living within the local pack's home range—could tell stories of encounters with wolves. Judging from what they said, many residents in the area have seen the wolf, and usually more than once. Although the speaker above conveys an idea of wolves that keep their distance from people and human settlements, and that seeing one is unexpected, too, the same speaker told us a story about an encounter with the predator. Wolves had visited the front yards of some of the other informants too, or had been seen in the immediate vicinity of their homes. One informant said a wolf regularly followed a "track" that went past her house. She had seen it through the kitchen window on more than one occasion. Yet, they all described their encounters with wolves as an extraordinary experience. For some, meeting a wolf was pure accident. Others emphasized that wild animals roaming around where they do not belong is wrong. But they all without exception saw these episodes as extraordinary—extraordinary because the very incarnation of the wild, the wolf, is "an alien" in the informants' own territory, in their own neighborhoods. The wolf's natural home is the wilderness. If it strays into farmyards—or near schools and day cares, for that matter—it challenges the mental boundary between wild and socialized. The social representation of the wolf is probably so firmly rooted in this divide that even the informants who spoke about regular encounters described the predator as essentially elusive and shy:

A: Jesus, we've been roaming around in these woods—you and me, and you and—

B: We've been at it twenty-five years in order to—

A: — to get a sight of that darned wolf.

B: We still haven't spotted it. We have seen traces of it, though.

A: Only [name redacted] has been lucky and caught a glimpse of it—

B: —for a couple of seconds before it vanished.

(Hikers)

Given this frame of understanding, it may not be surprising that encounters with wolves are like brushes with an envoy from another world paying us a brief visit, who disappears as soon as he gets a whiff of humans. Informants who had seen wolves emphasized the fleeting quality of these experiences, leaving the impression of an animal that only reveals itself in glimpses, as if materializing from nothing only to vanish again almost without trace:

> I have seen the wolf three times. The first time, it crossed the field, and then it ran up a slope. I ran after it. Then it stood still in the middle of the slope, and it turned and looked at me. [Speaking to another person in the group:] Then I felt like you said, when it looks you right in the eye—that look! And then it just continued up the incline. I went after it, and when I reached the slope, it was just gone—disappeared. The second time, I was on my way out the front door. At that time we had a lodger who had a cat, and the cat was out in the yard. Then I saw the wolf, standing there, in the gate. It was almost sitting, a sort of crouching, and peeked. But when it discovered me in the doorway, it was just gone—vanished! When I went to take a look at the tracks, it was obvious that it had rushed off. I saw it another time. I was out walking, and then it was just as if it evaporated from the road in front of me. So, when you ask what I associate with the wolf, I think about being shy, *incredibly* shy! [Affirmative exclamations from the others]
>
> (Neighbors group 1)

As we can see, the definite article is often used as a specific determiner in the phrase "the wolf." This may be the normal way of talking about large, wild animals—for example, saying, "I came across *the* moose when I went for a walk in the woods," despite never having seen *that* particular moose before. "I saw *the* dog in the park today," on the other hand, is a grammatically erroneous statement if the dog were a complete stranger. "Which dog?" the listener would probably ask. In everyday conversations, we just do not refer to dogs—or cats or mice, for that matter—in the definite form unless we are talking about specific individuals.

When it comes to the wolf, frequent use of the definite article seems to direct our attention to the species. "I've seen the wolf three times," however, does not necessarily imply that the same individual was observed on the three occasions but rather that the person who made the statement saw a specimen of the species on three separate occasions. The observed animal therefore performs the role of a sign, referring to something more general. It becomes a representative of the phenomenon *wolf*. Encounters between the wild and the socialized emerge as encounters between humans and individual animals. Even the informants who had observed two or more wolves together created in their tales the idea of the solitary wolf, or stray animal. The wanderer is per definition alone and on foreign soil. It plays the part of a visitor on a brief detour from its wild homeland and wild comrades:

> A: I think it's steering clear of the area round here, round the cabin here and—

B: We haven't had the good luck even to see traces of it round here—

A: True.

C: But there was that time two or three years ago, when it went down to the water here. A few people saw tracks. I saw the tracks too, and by then it had gone ashore in the cove over there.

B: But it's so rare—

C: But lone wolves do turn up, you know.

(Hikers)

"Wildness," then, constitutes a basic component of the informants' social representation of the wolf, associated in their stories with another core conception: the idea described above about wolves collectively, the "pack." The social proclivities of the conception: ability to organize family groups and work together—are a precondition of the species' strength and autonomy from other species, including humans. Group membership gives the wolf its fundamental independence and detachment from the human, and in that sense helps to sustain the mental boundary between the wild and the socialized. Thus, the representation of wildness reflects the idea of belonging to a group. A solitary wolf, a loner, might visit the socialized domain of humans once in a while. The pack, on the other hand, is inextricably bound to the wilderness.

THE PURE WOLF

Social anthropologist Mary Douglas (2002) explored the "symbolic order," the way cultures sort things that belong together into certain categories. This operation creates a system of distinctions that maintains order and brings reason to every cultural universe of meaning. Some things, however, do not fit any classification and are defined as foreign and impure. Things and phenomena easy to categorize are considered "pure." The "impure" penetrates the symbolic boundaries and threatens to create chaos. Douglas (1984, 2002) was interested in religious and symbolic relations to animals found in different cultures. In her account of the Lele people, she describes how the symbolic distinction between humans and animals constitutes the basic principle of the Lele cosmology. She explains how the Leles ascribe human features to animals that overstep these boundaries: "Most animals run away from the hunter and shun all human contact. Sometimes there are individual animals which, contrary to the habit of their kind, disregard the boundaries between humans and themselves. Such a deviation from characteristically animal behaviour shows them to be not entirely animal, but partly human" (1984: 24). Douglas sees notions of purity and impurity as deeply embedded in all cultures. All human collectives seek to impose order on the world by creating correspondence between their surroundings and established social thought patterns (Douglas 2002).

In the example from the Lele culture, animals are expected to avoid people. If they do not, their transgression puts the symbolic order at risk (Douglas 1984: 32). "Real" animals should never venture so close to people. Those that do correspond to the concept of impurity. The Lele deal with the danger of symbolic pollution by redefining and attributing human traits to the transgressing animals. In this way they remove the dirt, according to Douglas, and the categories remain unpolluted and intact. Could it be that the wolf threatens to pollute our cosmology in the same way? Is it regarded as a symbolic threat? The question we need to ask ourselves is whether the presence of the wolf undermines the distinction between wild and socialized. Douglas's identification of the impure, as something that by its very existence jeopardizes symbolic boundaries, refers to objects and phenomena that resist definitive categorization. Our interviewees in Trysil and Halden, however, doubtlessly define the wolf as "pure," since they unambiguously place it in the "wild" category. Incidents involving encounters with wolves are classified as extraordinary in a form of symbolic management of what is perceived as inconceivable or meaningless.

In the same way the Lele see particular qualities in animals that behave contrary to expectation, the informants in this study define boundary-crossing incidents involving wolves as exceptional encounters with lost loners or individuals. For our interviewees, contact between people and wolves represents what in Douglas's terminology is "anomalous" or "ambiguous." By defining the encounters as a deviant, the "socialized" and the "wild" are preserved as ordered and ordering categories: "When something is firmly classified as anomalous, the outline of the category to which it does not belong is clarified" (Douglas 2002: 53). But classifying encounters between wolves and people as anomalous does not necessarily mean our informants see the predator as polluting the socialized environment in a "negative" sense. Whether one thinks of the wolf's insertion into the socialized as something unclean or dirty will depend on one's view of the human society one belongs to. Should the human community be protected from what is wild, or is there room for "capsules" of the wild in human culture? There are several ways to approach anomalies, Douglas says. We can ignore them, express our disgust of them, or define them as dangerous or forbidden. But we can also respond positively to alien and deviant transgressions of established categories. The latter approach necessitates a collective "mopping-up" operation, where categories are redefined and space is found for the anomalous in a new system of thought (Douglas 2002).

A common thread in our interviews is the prominence of the wolf's wild nature. In defining carnivores that leave the wilderness and enter the socialized realm as anomalous—as special cases—informants confirm the wolf's definitive status as wild. There is a difference, however, between those who perceive the animal's presence as negative and the wild as a threat and those who wish to clear a space for the wild. It is, in other words, a question of what place the wild is accorded in modern society. We address how different ideas of the wolf connect to different understandings of the social environment later in this chapter. For the time being, suffice it to say that participants in the different focus groups share key notions of wildness and purity. If the wolf—in its essence—had been classified solely as a polluting agent, one would also have to question its wild nature. In that case, it could be classified as neither wild nor socialized and would threaten the validity of both categories. But this is not the case. A real wolf is inseparable from the wild and therefore pure. The same does not apply to what are known as "wolf hybrids."[4] They have no place in the wild or in the socialized world. By its very existence, the wolf hybrid undermines the structure of social thought. At risk is the established distinction between wild and domestic animals:

> That's the most dangerous wolf of all, if there's a dog mixed in it! It will have both the properties of a wild animal, plus it lacks its natural fear of people. That's definitely the most dangerous sort.
>
> (Hunters group 1)

Just as the wolf becomes the embodiment of all things wild, the dog is perhaps the strongest symbol of the domesticated. It is not called "man's best friend" without reason. But while dogs and wolves are subcategories of the same species in biological terms,[5] they represent opposite poles on the axis of conceivable relations between humans and animals. A hybrid, however, can be understood as neither a wild wolf nor a dog. It is simply nothing. Undefined, it emerges as "formless," in the way Douglas (2002) often describes symbolic impurities. Whatever lacks form cannot assume recognizable contours and thus becomes meaningless. One of our interviewees, a conservationist, believed that mixing dog and wolf was just as absurd as "mixing a moose with a horse, for example, if that were possible."

We asked our focus groups whether we should tolerate the presence of wolf hybrids in the Norwegian fauna. We wanted to encourage reflection and discussion on the topic of the wolf's essential nature and what, for instance, sets it apart from a dog. However, the conversations revealed what can best be described as a taboo: wolf hybrids appear extremely dangerous and need to be eliminated. On this point, there was virtually no disagreement. We asked exactly the same question in every interview and got the same response, whichever side of the carnivore conflict participants identified with. Of all our informants, only one had any doubts, but that person soon joined the rest of the group:

> Interviewer: Do you think it's right to kill [wolf hybrids]?
>
> A: Ah, that's a difficult question!

B: They should be put down.

C: Yes, they have different genes and might become friendly to humans.

D: Like I've said, I've worked with dogs with wolf blood, and that's something I never want to do again!

A: No.

D: Because you can't trust them.

A: No.

D: Because they haven't got that barrier, the wolf's natural fear of humans. And they're not responsive like dogs. (...) They're not nice things to have lurking around house corners, because they *will* lurk around house corners, of that I'm totally convinced ...

B: I would rather see them taken out, if that were possible. Because I think it would be like degenerating. We want pure wolf populations, with the natural instincts they're supposed to have. That's why I'm afraid that if we let litters like that live, I think it'll give arguments to people who say, "whatever it is you've got there"—they're not wolves at all. They're mutts! So personally, I'd prefer it if they were removed.

A: OK, that sounds reasonable enough, then.

(Neighbors group 2)

With the exception of this one informant, everyone in the group had the same opinion. Therefore, the topic was not much to discuss, but quite a few wanted to elaborate on their views on wolf hybrids. And the pollution of the idea of the pure wolf seems most unsettling:

Interviewer: What do you think of [culling of wolf hybrids]?

A: As I see it, that's no problem.

Interviewer: Why is that so?

A: Well, to keep [the wolf] pure.

B: It must be pure bred.

C: The gene material should be proper, you know.

(Conservationists)

Interviewer: Let's say you discovered some of these hybrids, mixtures of dog and wolf. How would you react?

A: Shoot 'em!

B: Yes, that's right, get rid of them. They're even more dangerous, because they've got even less respect for people.

A: If it happens in the wild, it will contaminate the wolf stock.

C: Yes, it would be destructive for the wolf.

(Mushers)

Words like "bastard," "pollution," "pure," and "dangerous" appeared spontaneously every time the topic was raised, leaving little doubt that wolf hybrids are classified as "impure," in Douglas's sense of the word. Unlike the stray wolf, which can also represent something out of the ordinary without necessarily provoking negative associations—and which, in all its solitude, confirms the wolf's purity and its proper place in the wilderness—the common notions of hybrids were uniformly judgmental and denying:

> Interviewer: So if wolf hybrids turned up—mixtures of wolves and dogs—how do you think we ought to react?
>
> A: Get rid of them!
>
> B: Yes, get rid of them!
>
> A: Goes without saying.
>
> (Farmers group 2)

This sense of unease connects particularly to unpredictability. If you mix wild and domestic, you do not know what will come out of it. The risk stems from the combination of two distinct categories, and the result can be an unnaturally socialized wild animal or, inversely, an aggressive and dangerous domestic animal. Hybrids will lack intact wolf instincts and will probably be incapable of surviving among wild wolves. Without the wolf's natural shyness, hybrids will gravitate toward people and therefore represent a danger. Several said telling a hybrid from a wolf, based on the exterior, would be almost impossible. The formless, fluid, and dangerous converge in the unpredictable behavior:

> A: It's not a wolf, though it might have some of the ... You don't know how it's going to react when you mix things together. You have no idea how dangerous it might be and what sort of genes it has. No idea.
>
> B: Semi-domesticated, then. [Others chuckle]
>
> A: Again, you have no idea. There can be differences between animals in the same litter. Impossible to know!
>
> (Farmers group 2)

Even though people are generally aware of dogs and wolves' common genetic ancestry, to most the animals remain two completely distinct "races" (the word used by our interviewees), clearly because the distinction between wild and domesticated species is a basic criterion for classifying animals. Apart from challenging the boundaries between what is wild and what is socialized—between wild and domesticated animals—the phenomenon of wolf hybrids may be confusing because it challenges the very concept of species. If social thought requires impenetrable barriers between different groups of animals—between cats, mice, lynx, horses, moose, dogs, and wolves—hybrids testify to the opposite: the permeability and fluidity of inter-species boundaries. At its most extreme and disquieting, it questions whatever is uniquely human. Perhaps the difference between humans and apes is equally as fluid. Can they mate and produce offspring? Has it been tried in our time? The questions are alien, the ideas preposterous, and the taboo unmistakable.

REPRESENTATIONS IN CONFLICT

We have concentrated so far on conceptions (of the wolf) whose nature are to provide a basis for other ideas, which are our subject here—the ideas that Moscovici and his colleagues call "peripheral" aspects of the social representation. These secondary notions tie into or relate to the basic assumptions, and disagreement over these thoughts is common. At this level, disputes occur between groups and between individuals. Controversies can also be articulated as individual ambivalence or as contradicting ideas held by the same person (Abric 1993). Whether the wolf belongs in Norwegian forests, whether it is perceived as threatening or threatened, and whether it represents a danger to people are questions participants in our study gave diverging answers to. There were disagreements not only between focus groups but also in discussions within them. Although the divisions between the pro- and anti-wolf attitudes emerge here, the views are usually nuanced and comprise doubt alongside ambiguity.

NATIVE OR OUT-OF-PLACE WOLF?

Though our informants appear to share a basic idea of what the wolf represents, they differ on one important issue. For some, the wolf does not belong in the forests that surround them. For others, it does, and they welcome its return and see the natural habitat of the wolf and surroundings in a completely different light from those who object to its presence. Whose territory are we dealing with? What kind of nature are we talking about, and is that nature in harmony with the wildness of the wolf? While farmers and hunters typically saw the natural physical environment as a landscape for sustainable *use*—as productive areas for logging, grazing, hunting and berry picking—the informants who expressed positive views on the presence of wolves saw this same environment as (more or less) untouched nature or "wilderness." To them, the forests of Trysil and Halden evoked associations of something authentic and original—something that was there before them and provided a reason for human existence, as well as a sense of continuity. Wilderness was thus represented both as an actual place and as essence. It stood in sharp contrast to the modern, overcrowded, and noisy civilization in which human bonds with nature are lost, and became the scene of potential reunion between human beings and their origins:

Interviewer: So what would you say are the best things about the area?

A: You mean right here?

B: Yes.

A: Well—there's nature and the silence. I am so fond of silence. I like it when I'm far away—in the forests.

(Neighbors group 1)

Several pro-wolf informants told us how much they appreciated the opportunity to roam "far" into the forests and "immerse" themselves in nature. Lots of space and silence are qualities they highlight in reference to their own neighborhoods. In the pristine wilderness, at a safe distance from the stress and hassle of human society, they find calm and regain a sense of balance. The wilderness becomes an arena where humans can restore bonds to what has been lost—what was before us—and gives meaning and continuity to our existence:

A: To experience something so authentic, in this [modern] society of ours—to me, that's incredible—but also a vital necessity! Everything is becoming so artificial. Things keep disappearing and disappearing. So, to be able to (...) be in touch with something so—it must have been like that for an eternity! [Affirmative mumbling from others in the group.] You can sit down and feel silence, a calmness, but also a feeling that surpasses all of that. For me, that is something that simply makes me feel at home in the world, unlike what I feel in other places.

Interviewer: But do you think the presence of wolves in the area contributes to those feelings?

A: Absolutely, yes, yes, yes!

(Neighbors group 2)

In surroundings like these, the wolves have an obvious place. The sense of being enveloped by wild nature in no way conflicts with the idea of the wild, majestic wolf. On the contrary, several informants believed the natural environment in eastern Norway is admirably suited to the large carnivore. One conservationist put it like this: "When we talk about areas near the border [with Sweden], I've said many times to myself, this must be a marvelous place to be a wolf—the terrain ... yes, everything." By its very presence, the carnivore helps create this sense of the primordial. It becomes living proof there are still areas that have escaped the human compulsion to socialize the wilderness and turn all that is natural into something artificial. It becomes a symbol of permanence and resistance—a symbol of survival of nature:

I've seen some of these animals myself occasionally. And there's this aura of eternity, you know, and that's what's disappearing elsewhere. We are losing it. Everything else is gone already. It's just credit cards and pin codes and then there's nothing else. For me, it's ... a substitute. And that's because it becomes perfect; it becomes whole. It's not only compartments of stuff.

(Neighbors group 1)

With the wolf in the forests, nature is complete, a harmonic whole. Without the wolf, the same forest is an incomplete piece of nature. Opinions like these find strongest expression in groups where participants have a pronounced positive view of the wolf's presence (conservationists, for example), but participants in some of the mixed groups also raised the issue. Nevertheless, nothing that emerged in the conversations suggested these informants believe unreservedly that the wolf belongs in their own neck of the wood, or in Norway as a whole. Such a view would have to be actively justified, and to some extent this was a subject of negotiation within the groups:

Interviewer: You used the word "primordial" about the wolf.

A: Yeah, I see it as—it's from way back [laughs], like it was always there, from a long, long time ago [laughs].

...

B: But they had pit traps for the wolf too, in the forests around here, to get rid of it. So—

A: True, they hunted it back then.

B: But those were completely different times. You can't compare then and now—two different situations.

(Neighbors group 2)

All the same, informants vacillated sufficiently for us to conclude that the representation of the wolf in harmony with its surroundings is not as entrenched in social thought as, for instance, the idea of the wild wolf. Whenever the former notion was invoked, it was inextricably linked to the idea of the natural environment as wilderness. Natural, wild wolves are in harmony with a natural, non-socialized environment.

The groups of predominantly anti-wolf informants, that is, hunters and farmers, saw the wolf as an "intruder." Participants in some of the mixed groups also doubted whether the wolf's presence was beneficial. The latter were willing to discuss the matter and less adamant in their opinions than the hunters and farmers, possibly because their groups included a range of different views. The skeptics perceive the local landscape—and most of the landscape in Norway—primarily as a socialized area, uncultivated but productive land. This land is essentially a cultural landscape where forestry, grazing, hunting, and fishing have a rightful place. If there is to be meaningful human activity on the land, there needs to be active stewardship. And using it, like it has been used for centuries, is the best form of stewardship:

> What scares me about the large carnivores is that the land will not be used. Then it will just become overgrown, and we will have the forest right up to [our doorstep]. That's exactly what we don't want! We want it to be an open landscape (...) that is used.
>
> (Farmers group 2)

Just as the idea of what is lost and what once was motivates the notion of wilderness, the image of the used and productive land evokes a sense of continuity, the past, and a legacy from earlier generations. Evidence of our ancestors' struggle to tame the wilderness is what places meaning on the physical environment, and these traces must be preserved, not as a museum but by active use. The cultural landscape must be protected against spontaneous reforestation, and livestock and huntable game must be shielded from carnivores. Seen from this perspective, the wolf will clearly not be considered a natural part of the environment. Wild wolves have no place in a cultural landscape. This understanding of the land clashes with the representation of the real wolf. Social thought about the carnivore in this case collides with the perception of the landscape, which was corroborated in the interview situation by statements to the effect that wilderness and pristine nature no longer exist in Norway. "It won't turn into wilderness just because there are wolves there," one of the mushers commented. A hunter put it like this:

> I think if we accept large carnivores in Norway, which will obviously cause problems, we need to have them in places where people won't be affected. Clearly, the number of large areas of wilderness is declining. In my opinion, there shouldn't be any wolves in Norway. There are people everywhere, and Norway's too small for wolves in that way. We haven't large enough wild areas.[6]
>
> (Hunters group 1)

Those with clear reservations about the carnivore told us people and wolves just don't mix, or, as one farmer put it, "Given what people want to make of their lives and the natural environment and, I nearly said, creation in general, wolves and people don't go together!" The wild wolf should keep to large, unpopulated areas, like the wide expanses in Canada or Russia, or even in Sweden. Norway is seen as a country with an evenly dispersed population, where nothing is left to wildness. In short, there is simply not enough space for wolves:

> Wolves can stay in Siberia, where nobody lives. I say people and wolves don't belong together! (...) Unlike in Sweden, our rural areas are populated. That's what we've voted for, and there's several generations of agreement in parliament to have settlements throughout the land. (...) And that means we can't have wolves here, because the two don't go together. But they can in Sweden. But then we'd have to build wolf fences along that border, so we can manage

it together. So in the vicinity of the border, the Swedes, you might say, need to take Norway's settlement patterns into account.[7]

(Farmers group 1)

When the "wild" and "autonomous" predator suddenly shows up in socialized areas, behaves as if it owns the land, and helps itself to game and livestock, it is seen as meaningless. It represents, in Douglas's (2002) terminology, a form of pollution. Wolves in the used, productive landscape blur the distinction between wild and socialized. Consequently, the animals that have settled in Norway cannot be understood as "real" wolves. A key point, then, is to distinguish between the social representation of the wolf as a species and notions of the animals currently living in Norway.

REAL WOLF?

Wolf skeptics' perception of the Scandinavian wolf population conflicts with the representation of the real wolf. The wolf as such can be a fascinating, intelligent, and beautiful wild animal, but these ideas connect to assumptions about the animal's natural environment and clash with the image of the physical landscape in Norway. Thus, the local wolf seems "unnatural." Accordingly, several informants highlighted unnatural characteristics of the predator. First, informants emphasized that DNA tests have revealed that the current Norwegian-Swedish wolf population does not descend from the animals that lived on the Scandinavian Peninsula from the Ice Age until around 1970. The newcomers have wandered from Finland or Russia, which has a very large population according to biologists—perhaps as many as 300,000 wolves "There's a huge number of wolves in the world at the moment, as far as I understand. It's not a problem," a farmer said. Since the wolves are not "native" Norwegians, and since the wolf is not a vulnerable species globally, many found the government's decision to protect the species in Norway difficult to understand. As long as the current Scandinavian wolf population is not considered a natural part of the Norwegian fauna, some see the authorities and conservationists' argument that the wolf contributes to increased biodiversity as essentially meaningless.

Some informants cast further doubt on the wolf's natural association with Norway, referring to stories about the secret introduction of wolves, allegedly organized by what they saw as extremists in the environmental movement. (We examine these stories and their social function in chapter 7.) Others hinted at the involvement of the government and biologists. The image of the wolf as intruder thus acquired even sharper contours:

There were these two professors at Uppsala University who made a statement about the first wolves and how they reacted to the wildlife fences along the big motorways in Sweden and how they moved in relation to the farms there. They had clearly been in captivity, the professors said at the time. So nobody really believes they weren't brought in and released, the first ones that came here.

(Farmers group 2)

Others used words like "sick" and "hardly viable" about the Norwegian wolf. References were again made to biological research and DNA tests revealing the extremely high rate of inbreeding in the Scandinavian wolf population (see, e.g., Liberg et al. 2005). And, others pointed out, the population is beset by physical problems like scabies and a short tail. According to some, saving such a defective population is pointless:

> A: No, Gulliver [a deceased alpha male in the local pack] was full brother to [the alpha female in the same pack]. So it was a bad strain in the one born up there.
>
> B: Because they were full siblings?
>
> A: Because they were full siblings, and they were even supposed to be siblings of the same generation. It's completely unlikely that they would manage to establish a [reproductive] pair, but—
>
> (...)
>
> C: The pack we've got around here, it's apparently the most inbred of all.
>
> B: But what you're saying now, as far as I can see, is that the pack doesn't really have the right to live.
>
> (Farmers group 1)

Finally, some interviewees questioned the racial purity of the Scandinavian wolf. Some of the more skeptical highlighted what they considered unnatural behavior, noting in particular the wolves' curiosity and boldness in venturing near humans and residential areas. What scientists and others term the "Scandinavian wolf" these informants see as an animal with a suspicious demeanor and lacking the usual shyness of wild wolves. In short, it behaves more like a dog. Accordingly, some believe many of the alleged wolves are actually hybrids. On several occasions thoughts were aired about the cub that escaped when the hybrid litter in the Moss pack was culled in 2009 (see, e.g., Andersen et al. 2003), which could have mated with other wolves and produced "infected" offspring. There are reasons to suspect, some said, that the Scandinavian wolf population originally stemmed from animals bred and introduced by environmental activists:

> I personally don't think the government is being completely frank. Because that pack in Moss ... it was the first pack where there was talk about bastards— also because of the way they acted: they were bold, ventured near houses and attacked dogs and the like, not like a normal wolf. I know someone who lives

in the area, and they saw a lot of vehicles—suspicious-looking vehicles, with these enormous cages in the back which they couldn't see into—and they never got any answers from the people about what they were up to when they stopped them. It started early—it's years and years ago—and they never explained what they were doing. But they realized eventually that—or they thought—that it had something to do with releasing [wolves into the wild]. And then there was this business about Moss and the other stuff came to light. So if that wolf mated with a dog, and they say that [was why] there were bastards, I'm not so sure about that.

(Hunters group II)

Therefore, a resolution to the conflict between the social representation of wolves and the idea of the socialized landscape is to construe the animal as something other than a real wolf: it bears all the marks of human contamination and is no longer a wild animal. The challenge to the fundamental symbolic divide between wild and socialized is thus neutralized. As we have seen, the notion of the wolf as a species—that is, the representation of the "real" wolf—is not affected by ideas of impurity. Symbolically, the real wolf does not emerge as polluted by humans. But real wolves simply do not exist in Norway. The animal living in our forests is understood as something else entirely: at worst a hybrid that pollutes not only the socialized landscape but also the very idea of the wild. The Norwegian wolf in this sense is doubly unwanted by its adversaries.

However, as evident from the interviews, not all wolf opponents support theories of clandestine introduction and ideas of impurity. In the hunter and farmer groups, one or two participants usually expressed these opinions, but the others were less outspoken. This could be interpreted as a form of tacit agreement if it were not for some participants expressing doubt through sounds and body language and sometimes verbally challenging such views:

A: Frankly, we have no use for wolves at all.

Interviewer 1: Lot of people would agree with you there.

A: First of all, the wolf was exterminated. It has absolutely no genetic connections to the area, so in fact it's illegal.

Interviewer 2: How do you mean illegal?

A: It's been released. You're not allowed to release animals.

Interviewer 2: No, that is true, but—

B: [Taking issue with A:] No, but no one has ever managed to prove it.

A: They have too!

B: Hang on, I watched this program on Swedish TV about all this stuff. They took DNA samples and whatever from the whole bunch. They couldn't prove they hadn't come from Finland or that they hadn't wandered here on their own. They just couldn't prove it.

C: No, but they can't disprove it either, to put it like that. There are things suggesting they were introduced.

B: That's an extreme enemy of the wolf [talking]—

C: No!

B: Well, that's what I think, anyway.

C: No!

(Hunters group 1)

Skeptics usually direct their opposition against people—*other* people. The subject of this chapter, however, is the social representation of the wolf. Although the status of the wolf as a symbol of urban penetration into rural Norway is obviously part of this representation (see chapters 3, 4, and 7), two factors indicate certain differences between these ideas and the core assumptions about the wolf.

First, a core component in the carnivore conflict relates to the role of local people in relation to their natural environment. Are they autonomous stewards of wildlife and landscapes, or are they merely cogs in a government machine? Who are really threatened, people or wolves? In other words, we are dealing with representations not only of the wolf and of the landscape but also with representations of people. Second, as we have seen, assumptions about "unnatural" wolf populations are not necessarily seen as obvious truths. They are the subjects of negotiations, also in anti-wolf circles. Unlike the implicit and general notions of the wolf's traits discussed earlier, the thoughts expressed here seem more like arguments in need of justification. Participants often argue by referring to what they take as scientific assessments and therefore irrefutable statements. For example, they called upon statements by "professors" and referred to "genetic research" on the origin and internal kinship of the Scandinavian wolf population.

THREATENED OR THREATENING WOLF?

People who see the wolf as a natural part of the Norwegian fauna also use biological research to legitimize their views. They referred on several occasions to scientific reports dealing with the increasing vulnerability of the isolated Scandinavian population, which ultimately could be threatened by extinction without an infusion of "fresh blood" from new immigrants. Friends of the wolf refer to the same scientific information on inbreeding and vulnerability as the skeptics do, but use it to emphasize the need for careful stewardship of what in their view is a threatened species:

A: The number of wolves can't be much lower than it is already, because then—

B: No, it's at rock bottom.

A: —and then they will just disappear.

C: They're all related.

B: Mmmm, true.

A: Closely related, too!

C: Like the ones [in the vicinity] that were tagged, they were littermates, [that is,] not littermates, but they came from the same place.

Interviewer: Mmmm, yeah, there are many of those animals that are much more closely related than normal siblings, for example.

C: Absolutely. It's not really good in the long run.

B: No, it's not good at all! That alone tells you there are too few; they're unable to sustain a normal, healthy line—with a few fresh genes every now and then.

(Conservationists)

These informants make no distinction between the wolf as a species and the variant found in Norway today, whether or not it originated in Russia. In their opinion, the wolf enriches Norwegian nature. It increases biodiversity, they say, echoing the reasoning of the authorities and biologists. One conservationist said: "Species diversity is important in my opinion, and humans are really a part of a whole. I don't think humans should sort of think they're better than ... all the other animals and plants. We're a part of a whole, and that's why I think we need to preserve the animals we have here." Within this frame of understanding, what threaten the wolf are primarily humans and their interference in the national environment. Ideas of biological diversity and threatened or vulnerable nature, represented in this case by carnivores, correspond to a specific notion of people's role in nature and their relation to the animals that live there. Here, the wolf is not the intruder; people encroaching on the wolf's domain are intruders. "I would go as far as to say that the last ten or twenty years of development is what makes the conflict," said a musher. "When we start colonizing their territory—it goes without saying, we're building on what essentially is their territory."

Wolf supporters see both carnivores and people as parts of a larger whole, as part of nature. Nonetheless, they positioned themselves consistently outside of nature when they spoke of wolves and described the natural surroundings as something humans should interfere with as little as possible. Thus, they take issue with the government's attempts to control the wolves' dispersion through the establishment of a special management zone:

A: What I think is that the wolf should be where it belongs naturally. People can't just draw a boundary and decide that's where the wolf can live. A wolf can't see any boundaries, and when a regional [large carnivore management board] can suggest that in areas that's under extra pressure from wolves, we should just capture a pack and move it [elsewhere] to spread the pressure, you

start wondering—totally ruins my confidence in the [regional boards], I have
to say.

B: There ought to be a committee for capturing us too, you know, and spread
us evenly.

(Neighbors group 1)

Thus, the wild animals belong in a pristine nature. The term "pristine" here refers to everything that humans have not been able to influence. The wilderness
and wild animals are characterized by an absence of human socialization. Systematic attempts to adapt the natural environment to human activity imperil
both the wild animals and their habitats. In this way, the wolf is perceived to
be threatened rather than as a threat. Fitting wolves with GPS or radio collars
is another example of detrimental interference or harmful socialization of the
wild:

A: It's one thing if the hunters maybe shoot a wolf every blue moon, but I
think [researcher's name redacted] and those guys, they were intent on tagging come hell and high water. That wolf bitch they caught and tagged not
long ago, it was almost a drug addict. [Several laugh a little]

Interviewer: How do you mean?

A: They were chased, then they were darted several times, and [then they were
manhandled]—they *pestered* the animals pure and simple.

(Conservationists)

However, wolf supporters are not against all forms of human activity in the
wild animals' territory. On the contrary, these informants often actively use the
forests for hiking, fishing, and picking berries, maybe even for hunting or dog
sledding. As we saw, they perceive the wilderness as authentic and genuine—a
source of recreation for modern people. But they also expose the idea of a fundamental and normative distinction between what they describe as "pristine
nature" and human enterprise. People should be allowed to enjoy the natural
environment as long as they respect it and do not affect it in any perceptible way.
People cannot act as if they are superior to animals.

For these informants, the wolf represents neither a threat to livelihoods nor
a competitor for resources. Passionate hunters, on the other hand, fear smaller
yields because the surplus of especially moose and roe deer is lost to the carnivore. They are also fearful for their hunting dogs, which are vulnerable to wolf
attacks. Landowners stand to lose income from selling fewer hunting permits.
For the farmers, the wolf is a direct threat to livestock. It reduces the available
grazing land and over the longer term is perceived as a threat to livestock production based on rough grazing. In the focus groups representing hunting and
farming interests, the wolf was consequently described not only as an intruder

but also as a competitor and a threat to people's livelihoods and way of life. These groups typically situate themselves *in* nature when discussing carnivore issues. They interact in various ways with the surrounding landscape and the animals living in it. Theirs is a utilitarian relationship to the physical environment in terms of farming, grazing, forestry, and hunting. Most hunters acknowledged that hunting is mainly a hobby and that providing food is not the primary motivation in our day and age. Nevertheless, hunters do in fact harvest nature's bounty. In this manner, hunters interact physically with the natural environment, just like farmers and foresters.

Government policy on carnivores in Norway, preventive measures, and compensation systems are mainly directed at the loss of grazing sheep. None of the informants included in the analysis of social representations kept sheep at the time of the interviews.[8] Nevertheless, many expressed solidarity with the sheep farmers and pointed to the effect of carnivores on the whole industry—not just loss of livestock but also farmers' identity. They were noticeably disappointed with the government's approach, claiming "society at large" had turned its back on farming and the rural cultural heritage. The wolf competes with not only sheep farmers for the right to use the traditional grazing but also landowners for the right to the game. A farmer asked, "Who shall have the right to live here, the wolf or us?"

> [Farmers in Østerdalen] earn their living from their grass and livestock put out to graze on the land. Without that, they wouldn't have two pennies to rub together. That's when you get the conflicts. We can put up with bears passing through and clawing a couple of moose, you know, and no one gets richer or poorer for that reason. That's OK, to a point. But when a whole pack of wolves ... returns on a regular basis and destroys—those [farmer-landowners in Østerdalen] may have to halve their [hunting] quotas. And all this has happened in the past ten years!
>
> (Farmers group 2)

However, the image of the wolf as a food competitor was strongest in relation to the hunters. In all of the focus groups and among wolf supporters and skeptics, the wolf was depicted as the hunters' rival:

> A: Obviously I know there are many hunters who think the wolf doesn't belong here.
>
> B: It's a rival!
>
> A: Yes, it's a rival.
>
> B: Quite simply.
>
> A: Yes, that seems to be what most people think.
>
> (Conservationists)

The hunter and the wolf are both predators. They hunt the same prey. The more moose and roe deer the wolf makes off with, the fewer are left for the hunter. Not only are the hunters' dogs and prey threatened but, as numerous informants emphasized, so is a way of life. One hunter exclaimed in frustration, "Forcing predators on people really affects the quality of life; that's just plain wrong, in my view. (...) That people can't enjoy their hobby anymore!"

Farmers were concerned about the harm wolves could inflict on the hunters' lifestyle and interests. That the wolf limits hunters' enjoyment of life and recreational activities is perceived as valid an argument against carnivores as, for instance, loss of livestock. But farmers' tendency to highlight hunting interests may also be a reciprocal message of support between two groups fighting the same battle. In their campaign against the wolf, hunters highlight loss of livestock and the decline in farming. Similarly, farmers highlight hunters' interests in the debate. Thus, both parties can claim to argue on behalf of not only themselves but also entire communities. As we discussed in chapter 3, the carnivore conflict has helped forge a symbolic alliance not only between local hunters and farmers, but also with landowners. Insofar as skeptics see the wolf as a threat, and not as threatened, their idea of the part humans play—and should play—in the natural environment also differs from that of the wolf supporters. As explained, the landscape for them becomes meaningful when it is used and useful. Since they actively utilize the physical environment for what they see as highly meaningful purposes, hunters, farmers, and landowners consider themselves "stewards" of the land. In their eyes, the government has demonstrated a total inability to defend resource-based industries and traditional harvesting. First, the cultural landscape is reverting to scrubland, caused partly by constraints on the use of uncultivated land in areas with large carnivores. Second, strict regulations and protection of carnivores have removed local people's right and opportunity to protect domestic animals and game.

We have described how the return of the wolf spurred the formation of new alliances across old social cleavages. The wolf became a symbol of a common external enemy, an urban intruder, but it also represents another symbolic threat. Some informants emphasized that if local knowledge is not used, the cultural landscape will deteriorate and animals will suffer. Apart from threatening grazing land, hunters' dogs, and game, the wolf also threatens hunters and farmers' self-perception as knowledgeable stewards of the land. As local people with local knowledge who uphold local practices in local surroundings, they see themselves as indispensable parts of nature's balancing act. An example is the way hunters use the notion of "predator control" to describe how they care for the local wildlife:

> We don't hunt only for food; we practice predator control. It has always been very important for hunters to kill predators, just as much as the game that

gives us meat. Controlling predator numbers has been important since the Ice Age. While the lynx mainly takes roe deer, it can also take hare and forest fowl. I actually see it as a duty to kill some of the large carnivores in order to maintain the balance of nature. It wouldn't be right to hunt only the animals we can eat.

(Hunters group 2)

Hunters and farmers wanted above all a decisive say in the management of wildlife populations, which also entails shooting, or "taking out," animals they consider harmful. Most said they were ready to tolerate a certain number of carnivores, including wolves, but only if they were allowed to eliminate nuisance individuals:

A: The authorities, at the very least they ought to let you remove the worst pests. So, the part of the country with the most bears, that's where we live. [Addressing another person in the group:] Maybe there are about ten different ones in the summer season that eat sheep. And if we could be allowed to remove the worst of them, there would be a bit more respect around for that there management system.

(Farmers group 1)

In the skeptics' opinion, limited opportunities to manage the local natural environment are to blame for a fauna that is not in balance. "There's hardly any roe deer left," one hunter said, speaking of the forests around his home. Because of the wolves, said a farmer who also happened to be a hunter, the moose "had vanished." Other informants made similar statements. The wild wolf's presence in areas defined as socialized does not only challenge the perception of the landscape and people's place in it. In the eyes of wolf opponents, the predator remains a threat to other wildlife populations because the central government prevents active local stewardship. Against this backdrop—and because the "new" Norwegian wolf did not originate in Scandinavia and shows signs of impurity—they argued that the carnivore also threatened biodiversity.

DANGEROUS WOLF?

Informants who were content with wolves in their area described the local fauna as "vital" and "intact." They claimed the carnivore had not affected the game populations to any noticeable extent. According to a participant in one of the neighbor groups, "There are as many roe deer here today as there were ten or twelve years ago. No one can say that numbers have declined!" Like his fellow wolf supporters, he wanted to show that the wolf did not represent a danger either to humans or wildlife. Friends and enemies of the wolf describe, in other words, two different realities. According to the former, the forests of eastern Norway are bursting with game. According to the latter, local fauna is under pressure from carnivores. The following leaves little doubt that the exchange of opinions by the opposing sides is part of a struggle for the right to define reality:

> I'm a keen hiker, as I've said, and I'm just as often on the Swedish side of the border as the Norwegian. And I meet people, all types, among them many hunters. (...) On one of my first outings after the wolf came back, I bumped into some Swedes during the moose hunting season. So I asked them how the hunt was going. "Brilliant," they said. "Full quota and no problems." "Thanks to the wolf," they said. "Thanks to the wolf?" I said. "Yes, 'cause it's driven all the moose," they said, "from the Norwegian side over to ours." "Well, that's fantastic," I said. "That's fantastic!" And it was fantastic. And they were pleased no end. Then I spoke with a group of Norwegian hunters, and they'd filled their quota as well. "And that's thanks to the wolf," they said. It was exactly the same, just in the opposite direction! [General laughter.] The wolf had run all the moose out of Sweden and into Norway. "Well, that's fantastic," I said. It was fantastic! And that's the problem, in a nutshell: "What's going to happen to all the animals?" But there are animals here. Tons of them!
>
> (Neighbors group 1)

Whether the wolf represented a threat not just to other animals but also to people was raised in all the groups, either by the interviewees or by the interviewers. The informants—wolf supporters as well as opponents—tend not to perceive the wolf as a physical threat to people. Of all informants, the strongest denial of claims that the wolf was a danger to people came from the two neighbor groups, which, as mentioned, consisted of people living within the home range of a wolf pack. They also had the most stories to tell about encounters with the animal. They agreed with the conservationists in denouncing claims that the wolf was a danger to people. Such allegations are a figment of the imagination, without hold in reality, they said:

> Interviewer 1: But don't people fear for their children ... have you heard anything about that?
>
> A: I can't say I've heard anything about that.
>
> B: Around here, the kids are waiting along the road for the bus.
>
> C: —waiting for the school bus. Nobody has ever been afraid of wolves.
>
> B: No, they really haven't.
>
> A: I've had wolves on my farm road, you know, and [my children] go there to take the bus. (...)
>
> Interviewer 2: Do you really never hear about people who are afraid of wolves?
>
> A: oh yes, we do! [Affirmative exclamations from the others]

B: But then that's media's fault. Whenever there has been an episode—somebody has seen a wolf in a built-up area (...) it attracts a lot of publicity in the newspapers. And then people who don't have any relation to wolves (...) and who aren't very interested, they only read the headlines and conclude that wolves are dangerous.

(Neighbors group 1)

Informants' ideas of fear and anxiety are colored above all by allegations put forward in the ongoing debate, the purpose of which is to describe reality so it corresponds to one's own opinions and attitudes. For friends of the wolf, the predator presents no danger to humans. Skeptics did not express any fear of the wolf either. The chances of being attacked by a wolf are infinitesimal, some of them insisted. But fear itself as a phenomenon and its effect on other people's quality of life did concern them:

My mother has been used to walking in the forests all her life. But since [the wolf came], she's never set her foot in the forest. As far as she's concerned ... and the older generation, it's got harder. They've got [an idea of the] wolf that's completely different from ours, probably. So she's not taken a step in the forest since that time, and that to me is really sad.

(Farmers group 2)

Children and seniors are particularly described as affected. "That it is a nuisance to us hunters, that's one thing, but we are *hunters*," a farmer/hunter exclaimed. "But we know some people are anxious for their kids—who live miles away from anyone—and they're too afraid to let their kids play outside around sunset, and especially after dark." Although their anxiety may not be justified, our informants believed it should be acknowledged, respected, and taken seriously in the debate about carnivores. Seniors who are afraid to visit the forests, children who are terrified of meeting a wolf on their way to school, and parents who fear for their kids' safety represent powerful reasons not to have wolves in Norway, seen from the viewpoint of wolf adversaries. Informant A in the excerpt below lives in the same area as those who believed no one in their neighborhood was worried for their children, but describes a different reality altogether:

Interviewer: Do you think a [harmonious] coexistence with the wolves is possible?

A: Well, it's something you just have to cope with. As long as they're out there, you just have to—

B: You can get used to anything.

A: Yes, but I know that there are many people who—maybe don't go out in the forest. You know, I'm lucky not to have that fear in me, but there is a group of cabins where I live and [the cabin owners] just don't go into the forest, because there are wolves there. And then of course you have to drive the kids to the bus. You get used to that too [said with an air of resignation].

B: But what's going on in the minds of the children whose dog has been devoured, you see, when the fear is there? We won't know the answer for a few years yet. Maybe the psyche of those children has been simply destroyed— quality of life of a child of ten, maybe destroyed seventy years' of its life—give or take. So you need spunk to think that's OK!

(Farmers group 2)

The big bad wolf of the fairy tales has no role in our interviewees' representations, whatever their standpoint. But those who expressed skepticism toward the predator implied that pictures of the dangerous wolf definitely exist in *other* people's consciousness, where it radiates fear and anxiety. Some use fear as an anti-carnivore argument since fear is assumed to impair other people's quality of life. Insofar as informants were willing to consider the wolf as an actual danger to children, comparisons with dogs were drawn. Everyone knows some dogs are not to be trusted, and we should expect some wolves to behave more aggressively than normal as well.

The notion that such behavior is "abnormal," including in wolves, must be seen in conjunction with the view that wolves visiting areas inhabited by humans is not "natural." The current Norwegian wolf cannot therefore be considered a true, authentic wolf. The wolf belongs in the wilderness, but its current habitat is not a wilderness but either productive land or populated areas. In that perspective, a wolf is about as dependable as an uncontrolled Rottweiler. But this purported anxiety is mostly a fabrication created by a sensation-hungry media, according to those in favor of wolves in their neighborhood, which brings us to what characterizes all the aspects of the wolf highlighted in the latter part of this chapter. They are all peripheral elements of the social representation of the wolf. They are topics of discussion and by no means taken for granted. In social thought about wolves, negotiations take place on whether the wolf belongs in Norway, whether it is a blessing or a threat to its surroundings, who should manage it, and whether it is dangerous.

SUMMING UP

While the wolf's place in the Norwegian forests is negotiated, its core qualities as an animal are taken for granted and featured in arguments for and against the wolf. In this context, the idea of the wolf's wild nature is key. Attitudes, negative and positive, are grounded in social conditions—in local and situated notions of animals, people, and nature. From the perspective of representation theory, explicit positive and negative opinions on the wolf link with ideas of the local physical environment, that is, to people's views on whether the carnivore belongs where they live. Many of those who describe themselves as wolf skeptics had nothing against the animal, per se, or the existence of carnivores in Norway. But, they said, the wolf did not belong in their neck of the woods—or in the productive and used Norwegian landscapes at all. Several spoke of the wolf in positive terms, as a fine animal as long as it lives in an environment more suited to its nature such as Canada, for instance, or Siberia.

People's opposition to wolves doubtlessly needs to be understood contextually, as opposition to local carnivores—animals out of place—rather than hostility to wolves as such. Expressions of attitudes are specific, then, to certain contexts and should be understood in such terms. Insofar as attitudes to carnivores connect to a local process of opinion formation, that people in urban settings are more likely to express a positive attitude to the wolf than those living in affected rural areas is not surprising. It is easy to be ecocentric on behalf of others. If the topic had been rats in apartment buildings, many people would probably have expressed themselves rather differently. The study of social representations corroborates interpretations that see the carnivore conflict as a reflection and manifestation of structural and cultural cleavages, as discussed in previous chapters. However, we are talking no longer about representations of the wolf but rather about expositions of human relationships. At the same time, if we focus exclusively on the cultural or economic cleavages exposed by the carnivore conflict, to identify people's relationship to the animal would be difficult. The shared representations described in the initial section of the chapter deal first and foremost with the characteristics of the wild animal, not with social relations.

Put differently, we can say social representations of the wolf consist of ideas at different levels. At one level, consensus and shared conceptions prevail; at another are diversity and negotiation. The two levels of social thought differ moreover in terms of content. At the diversity level, attitudes to the large carnivore also relate to antagonism between social classes, symbolic power, and cultural resistance, as well as to constructions of symbolic communities, as discussed in other chapters. Inasmuch as different attitudes are expressed in the current data, we need to see them as stories about people—oneself and others—about identity, power, and resistance. What are rumors of clandestine operations to reintroduce wolves, or myths about the laziness of farmers (more on those in chapter 7) if not arguments and allegations that primarily target something other than the large carnivore? In themselves, narratives like these can hardly be reasons to welcome the wolf back to Norway, or the opposite. Rather, they are arguments in a social conflict involving different segments of the population, in which many wolf opponents feel maligned by city people, scientists, and politicians and where the latter accuse the former of indolence, incompetence, and egotism. Seen from this angle, the wolf embodies discord, but again primarily in relations between people. If instead of concentrating on local people's opinions of carnivores we looked at their opinions of snowmobiling, we would very likely observe the same social constellations and mechanisms. Different social objects and phenomena reflect people's relationships to one another.

At another level, the carnivore issue is about people's relationship to animals—animals to which certain properties are attributed and valued differently. Some attributes are considered important while others are overlooked. Some are praised, others condemned. They are assessed, defined, sorted, and ranked. This is where we see the value of a different approach, one that can deal with more than conflict. By noting the ways social entities are described and characterized, rather than why people express different opinions on a particular matter, we reveal a level of social thought characterized by "common sense" and fundamental cultural categorizations of the physical and social world. At this level, a shared conception or mode of understanding prevails. For example, the wolf clearly rates highly in the social thought about animals. Irrespective of the feelings it provokes, it stands center stage in any discussion about carnivores.

In light of what we have described in this chapter, the social representation of the wolf can be said to contain a core element of consensus. Irrespective of opinions on the management regime for large carnivores, people think well of the wild wolf in its natural habitat. Consensus unravels, however, when people take a stance on whether the wolf belongs in the productive, used landscape, which for many is fundamentally different from "wilderness," and on whether the wolf residing in Norway today really is wild. This conclusion takes us in the same direction as the rest of our research: crucial dimensions of the wolf conflicts are not primarily about the animal itself; they are driven instead by other social cleavages with which the wolf's return and government policy on large carnivores can be associated. Importantly, many rural people perceive an unfair balance of power that benefits the "urban elites" while putting "ordinary country folk" at a disadvantage.

If one sees forests and mountain ranges as landscapes that are, and should be, marked by human enterprise, there may be no place for the wolf. To try to change perceptions of the wolf among opponents of the current management regime is hardly worth the effort. They already have a positive view of what they describe as real, wild wolves in a natural environment. Opposition to wolves arises in a conflict between social representations of the wolf and the representation of the land as productive, usable land, and these fundamental interpretations remain impervious to information from the government, scientists, or others. They are rooted in people's perception of themselves and the world, and any assertions challenging this perception will be dismantled and made to tally with existing perceptions, that is, the core of the representation. And as we made clear earlier in this chapter, this core is nonnegotiable.

If the goal is to get more people to accept the presence of wolves, to build on the undisputed core of the representation in an effort to extend this consensus might be sensible. One could argue, for instance, that the wolves already here—despite their Finnish-Russian genes—should be considered as real, wild animals, living in a habitat that is natural to them, although many would claim it is not a wilderness. The current management model, however, which to an extreme degree subjects the wolf and other large carnivores to human domination, effectively renders this approach impossible. Many see population control at the micro level, which requires GPS collaring and intensive monitoring of the largest number of animals as possible as a form of domestication. It whips the rug from under the argument that could have been deployed to encourage more people to accept the current Norwegian wolf as a representative of the "real wolf."

In addition, the extreme monitoring and control could induce people who are positive to wolves in Norway to turn against the government and its management system. They definitely want the wolf to be wild and are moreover worried about the ethical aspects of tagging and other manipulation of wild animals. Very few people would probably support total control of the wolf population by means of technology like GPS collars. The only reason for this type of management regime is to appease farming interests and to keep carnivore populations to a minimum without exterminating them. What it does, however, is intensify conflicts everywhere, especially in wolf areas, where livestock husbandry is limited. Farmer representatives are not happy either, because they object to presence of any wolves, and they perceive the "domestication" of wolves as a waste of resources and unethical to boot. Here, we see again how the core representation—the wolf must be *wild*—unites opponents and supporters of wolves in Norway.

NOTES

This chapter is a revised and extended version of Figari and Skogen 2011.

1. We use letters (A, B, C, etc.) to distinguish participants in the group interviews, but "A" and "B," for example, do not refer to the same persons in every quotation.
2. There was no permanent wolf presence in Trysil at the time. Some of the interviewees did tell us, however, they had seen tracks of wolves.
3. In Norwegian *revir*, a word also used figuratively to denote something that humans defend against intruders (e.g., a research topic or certified qualifications).
4. We use the term "wolf hybrid" for mixtures of dog and wolf resulting from spontaneous mating of a wild female wolf and a male domestic dog. Wolf hybrids live in the wild and are not reared by humans. We use the term "hybrid" because scientists and government agencies do. In 2000, the Directorate for Nature Management (now the Environment Agency) ordered the culling of a hybrid litter in the area around Moss, but one of the four hybrid cubs escaped. Their mother was allowed to live.

5. If a dog and a wolf mate, their offspring will be fertile. According to the traditional definition the two thus belong to the same species.

6. This may clash with a widespread notion—outside of Norway—that Norwegian nature is indeed pristine. But while it may be appealing and beautiful, it has been actively used for generations, for grazing, logging, and so on. And Norway does maintain dispersed settlement to a larger degree than most other northern countries, due to active regional policies in the entire postwar period.

7. The rural regions in Sweden do indeed have much fewer people than their Norwegian counterparts, due to a regional policy in the early postwar decades that was almost the exact opposite of the Norwegian one. Sweden encouraged centralization on a scale some would call brutal, to strengthen industrial development in urban areas.

8. As explained earlier, the chapters are based on different parts of our research, and the data for the individual chapters do not include every informant group. We *have* interviewed sheep farmers (see chapters 7 and 8), but they were not part of the subproject on which this chapter is based.

CHAPTER SIX

CONTESTED KNOWLEDGE

✻ ✻ ✻

In the previous chapter, we explained that the controversies over the return of wolves to Norway masks a wide-ranging consensus regarding the characteristics of the wolf as a species. The crucial questions are whether the wolf belongs in Norway today and whether the wolves here now can be seen as "natural" and "Norwegian." Consequently, people fight over most aspects of the wolf's return, and, as we shall now see, much of this controversy is embedded in a deeper conflict between science-based expert knowledge and lay knowledge based on practical experience. Much of the disagreement concerns population numbers and wolf behavior, especially related to humans. The nature of the interaction between wolves and people touches the core of the dispute: is the wolf in the wrong place or not, and is it really wild?

In communities with large carnivores in their vicinity, many people seem to be absolutely certain there are many more wolves, bears, lynx, or wolverines than biologists and managers claim. People also commonly say, in contrast to biologists, the wolves that have arrived behave "unnaturally." We will discuss this later. However, we also observe that people's attitudes toward large carnivores are to some extent related to their beliefs about population size. We do not see this as a simple causal relationship, but we observe a package of interrelated opinions: those who believe the populations are large (and larger than biologists claim) frequently feel more negatively toward protection of large carnivores than those who believe the populations are small (Bjerke et al. 1998; Bjerke et al. 2003). However, people who are positive toward large carnivores sometimes agree the populations are larger than the official numbers indicate.

It is a relatively common view in our study areas that the number of wolves is considerably higher than the official estimates, and quite a few contend that biologists' accounts of wolf behavior have been refuted one by one, when wolves have done what biologists said they would never do. Furthermore, there is a widespread skepticism toward the claim that wolves are not dangerous to humans, as biologists ostensibly contend. In areas where there are no wolves, but with bears, lynx, or wolverines, we find the same controversies over population sizes and what is considered natural behavior (see chapter 8). Thus, large carnivore conflicts, not least those involving the wolf, are also conflicts over what to consider valid knowledge.

As we have seen, very different attitudes toward wolves are found among people who live in wolf areas. Opposition against current protection of large carnivores is strong among hunters and others with firm ties to traditional land use and resource utilization, such as landowners and livestock farmers. However, considerable variation occurs within and between these groups, and some people without such ties to the land are also skeptical of large carnivores. An area with significant commonalities but also interesting and important differences within the "anti-predator front" is precisely the relationship to knowledge, in its scientific as well as lay forms. And—importantly—this is where some who support the wolf comeback take the same positions as their antagonists. As we have seen, some people in the wolf areas welcome the wolves back. We have already described a fundamental consensus on the nature of the true, wild wolf. We shall now see, perhaps surprisingly, considerable similarity between wolf opponents and wolf supporters concerning their relationship to scientific and lay knowledge.

On both sides of the conflict is a strained relationship with official knowledge about large carnivores and its source, biological research. We shall draw on examples from three of our study sites: Våler, Halden, and Trysil. For several years, Våler had a wolf pack nicknamed the Moss pack (after the nearby town) living largely within the municipality's borders. In 2000 and 2001, in the Moss pack's heyday, we conducted a study of the collaboration (or lack thereof) between important actors in the large carnivore field: managers, biologists, landowners, hunters, and a group we termed "wolf enthusiasts." The livestock farming in Våler was minimal at the time of our study, yet the conflict level was high. Within the municipality of Halden, where we conducted fieldwork in 2007 and 2008, are areas with the longest continuous wolf presence in modern times in Norway. The Dals Ed–Halden pack has operated on both sides of the border with Sweden since 1996. Halden is a medium-sized Norwegian town, and the wolves' presence affects a smaller part of the population than in the more typically rural communities we have studied. Still, considerable disagreement about wolves exists in Halden, and the relationship between official, science-based knowledge and local, lay knowledge emerged as an important topic in many interviews.

In the following discussion, we also draw on examples that are not about wolves. When we did our interviews in Trysil in 1999 and from 2007 to 2008, wolves were only on the outskirts of the municipality, as two packs had small parts of their territories within its borders. But since wolves were "everywhere" around them—in neighboring municipalities and not least across the border in Sweden—our Trysil informants were often very preoccupied with wolves. Quite a few feared the wolves were in the process of moving in, which is what has actually happened. Since 2010 the Slettås pack has stirred up controversy in not only the tiny hamlet of Slettås but indeed the whole of Trysil. However, just as important for our subject here (knowledge) is the presence of similar conflict lines involving the other large carnivore species. Even if hardly anybody wish to remove bears and lynx from Trysil, much disagreement exists on the population numbers and how these animals behave—and therefore about hunting quotas, culling, conflict mitigation, and so on. We shall try to shed light on some of the mechanisms that drive the conflict between local, informal knowledge and institutionalized, science-based knowledge. We will also attempt to place the knowledge dimension of the wolf conflict in a more general societal context.

POPULATION SIZE

Many people in our study areas spend a lot of time in the forest and have extensive knowledge about the land on which they live. They are used to making observations in nature, and they notice many things. Around wolf territories, they see wolf tracks, find wolf kills, and, from time to time, they see the wolf. In densely populated (relatively speaking) Våler, to see a wolf through the sitting room window was not too rare, and wolves sometimes threatened dogs chained or penned in the yard. During one of our first visits to Våler, we talked to a farmer who had just had a wolf walk across his field, and he showed us the tracks in the new snow. Immediately afterward we found fresh tracks along the roadside. Seeing wolves or wolf tracks was neither unusual nor difficult in Våler in 2000. The situation was much the same in Halden in 2007: quite a few of our informants had firsthand experience with wolves.

In tight social networks where information travels fast, to compile observations made in different places is easy. Thus, people can form opinions about the number of wolves in the area, and they can frequently put forward convincing arguments supporting their conclusions. Their reasoning is often detailed and complex and not something easily dismissed. Here is an example, taken from an interview with a hunter in Våler. Several others told this particular story as well—not only hunters but also some who were strongly involved in protecting the wolves:

> It would have helped if those people at the county[-level government environ-
> ment agency][1] could have a little more confidence in people, in the obser-
> vations we make. That's pretty much why people don't bother to report what
> they see. Because if there aren't any photos, or no tracks, if there is no snow,
> for example, then it seems like they don't pay any attention. Like this fall, there

was this guy who saw eight wolves coming by his moose post. And at exactly the same time another guy on the same team had a wolf five meters behind him. And that means that there were nine wolves, for sure, that they saw at the same time. And they couldn't see any collars on any of them [meaning they were not dogs]. They were pretty close, so I am sure they would have noticed that. (…) But the people from the county, they don't believe it, do they? If they could start believing what grown people tell them, and who are trustworthy, and so on. Then that would help a little.

(Hunter)

A government wildlife manager made the following comment:

In my opinion, we aren't so very secretive, at least not here at [the county-level environment agency]. But we cannot trust all the reports we receive, and mostly that concerns the number of animals. When we decide how much trust to put in a report, we need to find the person who saw something and talk to that person about it. But there are several people who have seen things—one has seen this, and the other has seen that—and at the same time, so it couldn't have been the same animals, and then there is four animals here and five there, and those are more than a kilometer away, so it couldn't have been the same ones, you know. We have had some problems like that, but a few phone calls and a bit of research and we soon find out that they are exactly the same animals, and then there are, let us say, five wolves that are observed with any degree of certainty. And then people get frustrated, of course, because they know that nine wolves have been reported. There are some problems there, definitely.

(Manager)

In 2000, when we visited Våler, the Norwegian-Swedish wolf research project Skandulv (see the introduction) estimated that the Moss pack consisted of at least five individuals. This was based on Skandulv's criteria for approval of observations, meaning authorized personnel must have verified tracks or made sight observations themselves. Population estimates based on these criteria are necessarily minimum numbers, as biologists and managers are now—as of 2016—quite careful to point out, and to add a maximum number, including uncertain observations, has become standard. However, during our fieldwork, biologists and managers usually presented the minimum numbers when talking about population size; for example, in the media and in public meetings (of which quite a few followed the first wolf arrival). As we can see, a considerable discrepancy exists between these numbers and the observations made by the moose hunters.

Hunters also present what they conceive as minimal numbers based on a compilation of observations. In the fall of 2000, their minimum was nine wolves, while the biologists said five. Like the biologists' estimates, the hunters' estimates come from observations made by persons deemed reliable—but these are other persons. Based on criteria different from those used by scientists, these people are trustworthy and known as experienced hunters and outdoorsmen, as steady, honest men who do not lie or exaggerate. Seen from the hunters' perspective, to disregard observations made by reliable people is absurd and humiliating. Furthermore, according to the hunters, doing so will give an incorrect picture of the wolf situation:

> Yes, they have taken a very arrogant position toward us. Fair enough, it must be a certain observation, but when you walk one and half kilometers into the forest to a small farmstead, no human tracks and no car tracks—and then suddenly you have tracks from two beasts with paws larger than apples, it's not a poodle you know. It's not a St. Bernhard either, because nobody would let it loose. It must be possible to use a bit of common sense.
>
> (Hunter)

Biologists counter that they cannot be certain about the observations' credibility, since they do not have the same knowledge about the experience and character of the sources. Scientific method requires all observations to be absolutely certain if they are to count (meaning the scientists are absolutely certain about them) and data must be collected according to specific procedures. Thus, population estimates will normally be minimum numbers or conservative calculations. One example can be found in the information brochure circulated to landowners from Skandulv and the Norwegian University of Life Sciences preceding an attempt to capture wolves and fit them with radio collars in the winter of 2001:

> A pack of eight wolves was observed during moose hunting, and a sighting of a single wolf was reported from another place at the same time. Skandulv today estimates that five individuals is the absolute minimum for the Moss pack. The low number is due to Skandulv's practice of only accepting tracks in snow and/or sightings during radio tracking as valid. This does not imply a dismissal of other observations, but reflects the need for a common standard that can be used in all areas with wolves.

On its website, the county-level agency reported several observations that in sum indicated a larger wolf population than Skandulv's estimate. However, this was done without calculating a specific population size.

Several people we interviewed admitted that scientists cannot say more than they are absolutely certain of, which means that numbers may be too low. Most accept that a scientific observation is something beyond just seeing something and talking about it. But their point is that this is irrelevant, since carnivore management should be based on the actual number of wolves, not on estimates from conservative scientific criteria. From the perspective of local hunters, that biologists and managers admitted there were more wolves than they could officially say was hardly any consolation. On the contrary, the management regime may appear somewhat absurd if biologists and managers are perceived as saying something like: "We know there are more wolves, but we can't admit it. We must manage based on population estimates that we know are too low."

So far, we have discussed what we may call different methodological approaches to population estimates: one lay and one scientific. As these approaches do actually result in different estimates, we could imagine this discrepancy as the core of what is often called a "data conflict": some people think there are many while others claim there are few. But it is not that simple. One very important factor here is that those who think there are many and those who think there are few are in different positions of power. Scientific knowledge has a dominant position in modern societies that is expressed in many ways, and we shall return to this. First, we shall look at what happens when these contrasting forms of knowledge meet face to face.

TRACKING

In Norway, particularly in the region where we conducted our studies, one arena where scientists and lay people meet is the large-scale snow tracking efforts undertaken each winter. Snow tracking is a staple in Norwegian wolf management because it provides an overview of the wolf population that is hardly possible in areas without snow. Luckily, most Norwegian wolves live in areas that have snow each winter, although in the southern part of the wolf range it may be variable. Tracking strains the resources of managers and biologists, and a huge amount of field time lies behind their population estimates. On the other hand, tracking is something most people, not only trained specialists, can do, distinguishing it from other techniques used to calculate population sizes that are often highly specialized, like DNA analysis.

The fact that most people can track has two important implications. First, it provides a foundation for alternative population estimates, based on local observations. Eager hunters and others who spend a lot of time outdoors regularly claim that their certain observations—in combination with solid knowledge of the land and extensive experience with wildlife—clearly show more lynx, for example, in their area than the official numbers indicate. Tracks are hard data, observed through an activity the hunters identify strongly with: walking in the backcountry, often several times a week for many years, and mostly in the same area. Such experience is claimed (no doubt rightly) to result in unique local knowledge and is contrasted with what is perceived as "desk knowledge" or lack of familiarity with local conditions. Second, the "democratic" nature of tracking also makes it well suited for collaboration between biologists, managers, and local people. In Hedmark there is now a long-standing tradition of large-scale tracking efforts to register lynx tracks on snow. This is organized as a collaboration between the county-level agency, biologists, and the Norwegian Association of Hunters and Anglers (NJFF) at the county level, who ensure that their local chapters participate. These chapters will again mobilize their members as volunteers all across the county. Some landowners and interested individuals also take part. However, hunters make up a significant majority of the volunteers, and the NJFF's central role underpins the hunters' crucial position, locally as well as at the regional level. Similar undertakings have been tried in other counties, with somewhat varied success.

The purpose of the collaboration, according to management agencies, scientists, and the NJFF, is of course to obtain better data so population numbers can be determined as accurately as possible. The most important species in these registration efforts is the lynx because it is hunted and hunting quotas based on population size are needed, though tracks from other species, including wolves, are also registered. Of particular importance for the NJFF is that the lynx quotas not be too small. More people in the field are basically assumed to lead to more track observations and thus contribute to an upward correction of the population estimate. For scientists and managers, to have as many observations as possible is obviously good.

An additional purpose is to improve the relationship between local hunters on one side and biologists and managers on the other and to minimize disagreement about population numbers. Over the years, a realization has emerged that the contact between different groups may itself erode some barriers and that a feeling of being taken seriously can alleviate the hunters' (and other local participants') experience of being excluded (Skogen 2003). An important element in the monitoring procedure is the verification of observations made by local volunteers. Personnel from the Norwegian Nature Inspectorate (Statens Naturoppsyn)[2] must check all observations. During the coordinated registration efforts, and on other occasions when observations are checked, those who made the observations have encountered experts representing the authorities face to face.

When we interviewed hunters, managers, biologists, and representatives of the NJFF in 2001 and 2002, many thought the collaboration had improved the relationship between "ordinary hunters" and managers/biologists and that there was more mutual confidence (Skogen 2003). This study had relatively few informants, but those we interviewed were central actors in the registration efforts. Even if several problems were identified, the dominant impression was one of goodwill and positive experiences. According to an experienced biologist we interviewed in 2002:

A: In the area where we have packs, two-thirds of the observations we check are not from wolves. In those situations, it is important to involve the person who made the observation and discuss it. And if we can't agree, it is a good thing that snow never lies, and we can follow the tracks until we do agree. If we think it is a dog, and the person who made the observation still thinks it is a wolf, well, fair enough, then we keep on going until we find, for example, the place where a car parked and let out the dog or into the farmyard and find the dog itself.

Interviewer: But does it ever happen that it really is a wolf, even if you doubted it at first?

A: Oh yes, that definitely happens!

Biologists also emphasize the scientific value of lay people's observations. Here is the experienced researcher once more:

Interviewer: Have these tracking efforts yielded information that you regard as scientifically valuable?

A: Yes, very much so. There are many methods you can use to register carnivore populations, and you will arrive at slightly different assessments depending on the method: weather conditions and the location of the animals on a specific day, and so on. So the more methods you use, the more accurate picture you get of what is really out there.

For biologists and managers, people's general trust of population estimates from conventional scientific methods would be a great advantage. One also gets the impression that they think this a likely outcome if lay people are involved in data collection. A manager said:

There are always rumors about wolves in areas where we don't have any confirmed observations. People are convinced that wolves have settled nearby. If no wolves are found during these organized tracking efforts, except where we already expect them to be of course, then maybe we can get rid of these rumors. So I have concluded that these collaborative efforts are important.

We interviewed a number of "ordinary" hunters about large carnivores and large carnivore management in roughly the same period (1999–2002). Not many negative statements were made in these interviews about the tracking efforts and the collaboration they depend on. In the early years of the new century, attitudes toward tracking and collaboration were generally positive. But in the interviews conducted in 2007, Trysil hunters spoke in a different tone. Many we interviewed had previously participated in tracking but not anymore. The reasons were manifold but mostly centered on an experience of the work as wasted: population estimates were not adjusted upward to a level hunters considered real, even if they participated and did their best. In their opinion, the registration program was rigid and incapable of absorbing local knowledge. They claimed transect lines (to be meticulously followed to see if tracks crossed them) were often drawn in the wrong places, not where local knowledge indicated lynx would be. Many hunters we talked to claimed certain knowledge about this, but as they saw it, their insights were not reflected in the plans for tracking. Their impression was that the organizers had no interest in local knowledge and were not prepared to integrate it in their planning:

> I did it a couple of years and walked these [transect lines], but I haven't done it the last few years. It's been two years since I last participated. The way this was handled—I was really annoyed, and it is not something I will use my weekend to do anymore. We were out searching, searching for lynx, and up in [place A] and also in [place B] there were tracks, and there were tracks here, and in [place C], and in addition they found tracks in [area X], a distance of sixty to seventy kilometers. But then they have to send specially appointed people to evaluate if we can recognize a lynx track at all. They barely made it [because of the great distances] that day. That's one thing. And then another thing—they came, checked the tracks, and then they told us that this was *one* lynx. "But how the hell can you know that?", I asked. Because they had backtracked it, he said. But even [famous Norwegian skier] couldn't have moved through the deep, soft snow from [place A], up to [place B], then down here, over to [place C], then down to [place D], and out into [area X], because it was snowing heavily. It's a lie, plain and simple. But they want it to look like there are no lynx. We found tracks from five, and they gave us just one.
>
> (Hunter)

Snow tracking provides good data and moreover reduces disagreement about population numbers through a method that requires local participation. But as we can see, the activity has not necessarily led to such an outcome. The hunter quoted above does not hesitate to accuse biologists of lying and having preconceived opinions on the number of lynx in a particular area. From such a perspective, participating in tracking is now pointless. Another participant in the same focus group added:

> One year there was this [transect line] in the southern part of our neighborhood here. It runs from the top over there to the road down here. And you are supposed to notice all animals that cross it, also lynx. So we walked from the top down to the road, and between the road and the river we found tracks from a lynx with two cubs; that is three lynx. They should not be counted in the report, because they hadn't crossed the [line]. Then I don't see the point anymore. OK, if you are looking for hare or forest fowl, then you need a line and calculate an average, I can see that, but we are talking about large predators here, and they are not so numerous that you need to do it that way.
>
> (Hunter)

The scientific method is seen as far too rigid. That lynx should not be counted just because they do not cross a transect line, when you know they are in the

area, is considered sheer nonsense. The whole point of the activity and the collaboration becomes obscured for those who see it this way. No wonder, then,
these hunters no longer want to participate.

Furthermore, the hunters are critical of the principle that all tracking must
take place in a single day, the rationale being to avoid counting the same animal
twice. But our informants claim that lynx often rest for a day or more, as has
been the case with known lynx during the tracking efforts. Accordingly, these
animals are not officially counted. In the hunters' experience, to include such
certain observations in the population estimates is impossible:

> You know, it's the lynx that we are most interested in, because that is the only
> license hunting we have in our area. And so we keep an eye on it all through
> the winter. It's a little frustrating: On the single tracking day we may be able
> to find tracks from only one animal, but we know that we have had a family
> group here, and that is critical for the hunting quota. And then it is very an-
> noying that only this single day is included in population estimate, when you
> know that the total population is larger than what is registered that day. When
> you spend time in the forest, you observe that the lynx can be stationary for a
> long time if it has killed a roe deer or something, and then it is very difficult to
> find if you rely on those [transect lines].
>
> (Hunter)

In several interviews, hunters complained that the quotas appeared to have been
set before the tracking. One episode in particular was recounted, where the requirement for allowing hunting (a certain minimum number of family groups)
was jacked up because "too many" lynx were found. In this case, the observations were accepted by those sent to check them, according to our informants,
but the threshold for setting a quota was changed:

> It was the same last year. To allow hunting here, they required a minimum of
> ten family groups in Hedmark. We found twelve, and then they changed that,
> then it had to be the average for the last three years. So there was no hunting,
> because they changed it. Then they wanted the average for three years. That's
> not right. Then there is no use.
>
> (Hunter)

Criticism of lynx tracking contains two main elements. The first concerns
the methods used: rigid transect lines and concentration of all effort within
narrow timeslots yield too low population numbers. This type of criticism could
have emerged at a conference for wildlife biologists. The other dimension of
criticism considers the tracking effort a mere masquerade. Regardless of what
is observed, tracking has no consequence on how the population is managed,
and quotas are not affected. In other words, the hunters imply that scientists'
agenda is not about accumulating reliable knowledge and that biologists act as

agents for a pro-carnivore alliance. Science is seen not as neutral and objective but rather as an expression of a political view entailing large carnivore protection. In such a context, hunters are unsurprisingly suspicious that methods have been chosen to systematically underestimate the number of lynx.

The strained relationship between managers/biologists and local outdoorspeople is even more evident concerning wolves. Many informants describe the fact that management agencies do not take reported wolf observations seriously as a comprehensive and enduring problem—a problem evident in Våler during our fieldwork. Several people told us they had been treated in a patronizing way and had experienced humiliating situations. Interestingly, these stories came from local wolf enthusiasts as well as hunters.

Those who have reported observations generally consider themselves experienced outdoorspeople and, most often, hunters. Some also enjoy such a reputation in their community, which may be an important part of their identity. They claim they are perfectly capable of identifying a wolf track and know if they have seen a wolf or a dog while moose hunting. So it can be an annoying, as well as humbling, experience to be told that what you have seen can be anything from a dog to a roe deer but hardly a wolf. This experience becomes even more serious if the person who gives that message lacks local legitimacy, is seen as having mainly "desk knowledge," and maybe even as a powerful antagonist in questions related to large carnivore policy. When observations are interpreted, the question is what type of knowledge should be trusted—local experience and common sense or the scientific method and abstract, academic knowledge. Most academics (i.e., biologists) who inspect tracks and evaluate observations have practical experience themselves, and frequently a lot of it, but many people are reluctant to accept this fact because the roles are quite dissimilar. While both parties may see themselves as experts, and may be recognized as such in specific (but different) social milieus, only one of them has a formal authority and contributes to the official population estimates:

> They reject it, that's what they do, and say that there aren't so many wolves here, but that doesn't add up when we are out in the forest and see what it looks like. Of course it is very difficult to prove these things, but we have seen it now with the snow, and we have people who have seen nine at the same time. And that doesn't fit well with the picture the managers say they have.
>
> (Hunter)

These situations can certainly be difficult to handle. The feeling that one's judgment and knowledge are doubted is unpleasant for anybody, particularly for those who have invested in their role as local experts. Local experts often have key roles in local wildlife management (on private properties and at the municipal level) and the community of local hunters. Therefore, a lack of confidence in the knowledge such people disseminate can be easily interpreted as a wholesale lack of confidence in much larger groups, namely those who trust the insights of local experts.

Among our informants, some also warmly welcome the wolf and would like to work with biologists and managers. Even if some have been accepted and allowed to participate in various activities, wolf supporters also report that they are not taken seriously, and their experiences seem strikingly similar to those of the hunters:

> No, it didn't suit the managers to hear about more wolves, because they had more than enough. They worked long hours because at the time they had the trouble with the hybrid cubs. And they had all the other wildlife in Østfold to manage. Moose and roe deer and birds and everything, and the wolf occupied them twenty-six hours a day! And then suddenly they were told that—then a layperson comes along and says that, "Hey, we have two wolves on the other side of the river that stick together; could it be a male and a young female?" It didn't suit them to have that information, and they didn't want the locals to know, so they wouldn't agree to that.
>
> (Wolf enthusiast)

> In connection with the tracking in Østfold organized by the County Governor, they have involved the landowners. Otherwise they haven't involved many people. When they tracked the hybrids, several of us volunteered because we would like to share our experience of the hybrid and the pack in Våler. We weren't allowed to participate. They wouldn't have any interference from private citizens. Quite simply.
>
> (Wolf enthusiast)

Here we see the same phenomenon we observed in hunters (who were not particularly fond of wolves): people who see themselves as experienced and competent feel snubbed and ridiculed, in this case, people who would actually like to have a close relationship with biologists and managers.

Our informants from the hunting community and those who are enthusiastic about wolves have interesting common features, which to some extent can probably explain what we see here: they spend a lot of time in the forest, and hunting or a strong interest in wildlife (often a combination of the two) are very important components in their lives. They have received recognition for their considerable amount of acquired experience and knowledge, which appears to be a significant part of what we may call their identity project. But compared to the biologists and managers, they remain amateurs who do not occupy positions that could lend official authority to their knowledge. Neither is their knowledge acquired in ways that lead to official certification in our type of society. Thus, the conflicts over knowledge are embedded in power relations, where scientific knowledge is always trumping other forms of knowledge in decisions on actual large carnivore policy. Consequently, those who represent lay, experience-based knowledge endure one defeat after another.

We do not intend to describe what has "really" happened in encounters between certified experts and the local variety. We are concerned with our informants' experience and interpretation of what happened—their understanding of the situation and the consequences this understanding leads to in terms of action. Regardless of whether one's aim is scientific understanding of a form of social interaction or identification of practical implications for policy and management, this is at the core of the matter. If people think they are being treated unfairly, this understanding underpins their attitudes and actions and therefore has practical consequences that cannot be disregarded.

WOLF BEHAVIOR

Disputes over observations and population numbers are not the only examples of knowledge conflicts surrounding the wolf. In several areas, some groups in the communities we studied (our hunter informants among those) stand firmly against biologists and managers. This relates to different aspects of wolf behavior, where biologists' claims are said to have been disproved one by one (by the wolf itself), but perhaps the primary question concerns the danger wolves pose to humans. Again, the point is not whether our informants have misunderstood the scientists or if they have registered that scientists may have revised their own views over time, but rather to show that people experience a basic knowledge conflict:

> It was the week before the school holiday, up at [place X], north of here. These two girls are waiting up there where the school bus turns around twice in twenty minutes, in the morning, picking up the elementary school kids first and then the junior high kids afterward. And when the driver returns to pick up junior high, then these girls are standing there, they are eighth and ninth grade, and they stand there talking at the roadside. When they enter the bus, when he shuts the door and starts moving, then the wolf rises from where it has been lurking. He was looking at them from across the road. And then you wonder why he was there in the first place. Because not even a magpie would stay so close to people if they were moving. It's five meters, not more. And it causes fear among people and an eerie atmosphere. And it was a fifty-year-old man who saw it, steady, churchgoer and all that. Never been involved in any nonsense. So he tells about this, and the newspapers pick up the story, but then the people at the County Governor says it isn't true. Then I wonder what is really going on.
>
> (Hunter)

Last year there were two or three wolves at [place Y]. Two of them came to the farm next to where I live. A single woman runs that farm. She saw with her own eyes one of the wolves jump a fence that was about [120 centimeters high]. Then the managers come along to have a look. And we were standing there and we could see, because the ground was muddy, the prints where one of them had jumped and where it had landed again. But this guy from the County Governor, he says no! Because there was a thirty-five-centimeter opening under the fence, and he said, it's not a problem for a wolf to crawl under. "No," we answered, "everyone can understand that, but here we have the paw prints that shows that he jumped this time. There he jumped, and there he landed." But no, it had crawled under.

(Hunter)

Because there are too many—well, perhaps not lies, but at least half-truths or not saying it like it really is. Because these animals, you know, they said they wouldn't do this and they wouldn't do that, but there isn't a single one of those things they haven't done. They wouldn't cross streams or bridges or jump fences or whatever it was, but why shouldn't they? I mean, if the food is on the other side of the fence. These animals are supposed to be so smart, maybe the smartest animals there are, and they have to get there somehow. So they must jump and eat the food on the other side. It's obvious and quite simple, really. So I mean, those people who feel they are in charge here, they only talk bullshit.

(Hunter)

These quotations illustrate what people feel when they are not believed and, as we can see, not only about population numbers. When it comes to wolf behavior, people have also seen how those who represent formalized knowledge doubt specific observations and eyewitness accounts, which creates and maintains mutual distrust. But we can learn something else from this: our informants think they are knowledgeable themselves. They do not see a great need for information from scientists about wolf behavior—at least not from sources that have been repeatedly discredited through *mis*information. Those who adhere to practical lay knowledge do not see their own insights as inferior—quite the contrary. No wonder, then, they are provoked when science-based academic knowledge is given so much more weight in institutions with power over wildlife management and conservation.

However, our informants are not alone. Many dismiss scientists' explanations. For example, one might almost say an industry has evolved around producing evidence to prove that wolves are dangerous to humans. This is done by studying local historic material, such as parish records, and digging out reports (especially from Russia) unacknowledged by Norwegian biologists (and probably by most other modern biologists). These reports recount wolf attacks on people, even from recent times. The efforts of the counter-expertise are often well known among those critical of current large carnivore policy, and their work is regularly disseminated through various websites. Books and pamphlets are also produced in both Norway and Sweden (mostly published privately but sometimes by small publishers). A network of wolf counter-experts is active in the two countries. Some are very adept writers and obviously invest a tremendous amount of work in their calling.

A significant factor here is obviously that biologists' refutation of the claim that wolves are dangerous may seem to have so many holes that most people will raise an eyebrow. A layperson may ask if the wolf's harmlessness is really substantiated by the lack of fatality records in sources from the nineteenth century or even further back. Maybe it is not so strange there are few reports about wolf attacks after the early 1800s: "We have had almost no wolves in Norway for more than a hundred years, so how could there be any attacks?" is an argument we have heard often. Here the scientific method clashes with everyday common sense. Very few of us organize our lives based on scientific principles. Arguments that come across as formalistic and abstract reinforce the impression of scientists as out-of-touch academics with nothing to offer in the real world. This paves the way for the counter-experts, who can tie into modes of understanding that people recognize and that we all resort to in our daily lives.

However, scientists and managers also acknowledge that many mistakes have been made in this area, especially in the early years of the wolf comeback. These mistakes were probably due in part to the particular intensity of the conflicts over wolves. As one biologist said in an interview, "We may have lulled ourselves into a notion of wolf research and wolf management as similar to any other large carnivore stuff." Even if other species also cause conflict, the wolf seems to hold a special position. Managers tend to blame media for escalating the conflict level through the incessant quest for sensation. They claim that media frequently misrepresent facts and stoke aggression, especially among wolf adversaries. Explaining an effort to counter this "misinformation" on its website, one representative of the County Governor in Østfold said:

> People believe more or less what suits them in relation to the opinion they have of an issue, and they select pieces of the truth that suit them at the moment. So I believe that one of the best tools we have is to try to be on the offensive all the time and inform people that so many wolves were seen here or there, and they did this or that.

(Manager)

In the early 2000s, our distinct impression was that biologists for their part found it difficult to relate to local input. One researcher interviewed in 2001 said:

We have discussed local observations at length. I think it is very challenging. It may only get worse if you start sifting reports and say, "OK, he saw eight, but this other guy saw nine, but he who saw eight is more trustworthy that he who saw nine." Start to qualify, and sorting people into categories based on whether they are liers or not, or have a more or less vivid imagination, that's a recipe for disaster. Obviously, it is particularly difficult here where we have so poor tracking conditions because of variable snow cover. It is better where they have these organized snow tracking events and many hunters take part.

(Biologist)

At the time, biologists and managers generally described local "experts," particularly those who did not want wolves, in negative terms. Many of them were labeled "self-appointed chiefs" of an arrogant macho type, who used the wolf as a tool to bolster their position among peers. They were said to pursue knowledge in a very selective way. Biologists and managers alike apparently wanted to keep such people at arm's length.

Biologists and managers referred to what they saw as local myths about wolves, and imaginative stories circulated among wolf opponents, stoking fear and general opposition to wolves in the neighborhood. A common story told in Østfold that several managers also knew was about children at a day care center who met a wolf in the forest. Told by a local informant:

This wasn't so long ago. It was near the village center here, in a day care right behind the school. They have this small forest that they use to play in, just beside the day care. One morning when they went over to their little forest, a wolf was sitting there. And there are houses all around, but there is this small forest patch down there. So why was the wolf there? If a kid had gone over there alone, no one can say what might have happened, and people don't want to take that chance, you know.

(Hunter)

Managers seemed quite familiar with such stories but were surprised these events had not been properly reported. For example, they were never notified of the day care episode but had heard about it through the grapevine, like anybody else. Managers took the lack of reporting as an indication that the stories were not true or were greatly exaggerated. They explained these rumors by pointing to the general tendency of "village gossip," so common in rural areas. They also believed quite a few local characters completely "took off" when the topic was large carnivores, so managers generally rejected the content of these stories. Stories about illegal killings of wolves were also doubted, although nobody denied that this might have happened. Managers seriously doubted local hunters would be able to shoot wolves even if they tried, or that they would manage to go undetected if they did. At the very least, managers were certain that the number of wolves killed was seriously exaggerated in these stories. They found it difficult to improve local knowledge about wolves as long as stories without foundation circulated and proved to be very resilient:

> There are all these really silly stories. If you use a bit of common sense, you will understand that it is not possible. But they are circulated very actively, and those who tell them or believe in them are not so concerned with what is true.
>
> (Manager)

The biologists had similar problems accepting what one described as "the propaganda against wolves that some people enjoy spreading out there in the outback." However, managers and biologists agreed that science could contribute to shooting down the most exaggerated wolf stories:

> Everybody claims that wolves are lurking outside their houses all the time. So it's a bit like it was up in Skiptvet [municipality in Østfold] the summer we had the hybrid cubs; there were wolves in all barns and yards and sandboxes, and I don't know what they didn't say. But then one of them got radio collared, and suddenly everything was silent. They hardly see a wolf anymore, and that's a bit strange, isn't it?
>
> (Manager)

Although some managers and biologists may still hold such views about local people, we can clearly see a trend toward involving local communities more in management. Extensive, coordinated tracking efforts are one example. Despite setbacks, the relationship between local hunters and biologists and managers has improved, not only from the tracking but also because many biologists seem to have changed their attitudes toward locals. They are more inclined to take local, informal knowledge seriously and to emphasize staying in touch with local people. This may pay off for at least two reasons: Lay people may possess valuable knowledge (valuable also to science), and keeping in touch with them can take some of the edge off (some) laypeople's critical attitude toward scientists.

CONTESTED KNOWLEDGE IN MODERN SOCIETIES

To better understand the conflicts over knowledge that play out in the large carnivore field, we need to take a closer look at the position of science and expert knowledge in modern societies. Many social science scholars see contestation of expert knowledge as characteristic of the times we live in, which have been labeled "late modernity" (Giddens 1991). There are several reasons for this, but according to a number of sociologists, a core point is that the cultivation of science and rationality in modernity strikes back and undermines the authority of science.[3] Granted, we are more dependent on expert knowledge than ever before, and our understanding of the world around us is increasingly based on knowledge originating in science and not from our own experience. But there are so many experts, and they often disagree, so it is possible to find competing ways to understand almost any phenomenon. More education and new and effective media have made it easier to compile alternative knowledge that challenges official knowledge on its own premises. One can almost choose the knowledge and the experts one needs: there will always be experts who criticize the dominant understanding of any phenomenon. Some also claim (e.g., Beck 1992, 1995) it has become easier to disclose connections between knowledge and power and to see that scientific knowledge is used to bolster certain economic or political interests.

Official expert knowledge, not least in the environmental field, has been challenged, which is often used as an example of its weakened position in our time (Beck 1995, 2000). The environmental movement and a growing environmental awareness in large segments of the population have no doubt driven the dominant (hegemonic)[4] knowledge on the defensive from the 1960s onward. Everything form nuclear power and pesticides to clear-cutting and draining marshes have at some point been declared harmless. Today, GMOs and electromagnetic radiation are examples of issues where official expert knowledge dismisses concern voiced by environmental organizations, critical scientists, and many lay-people. These are areas where the links between knowledge, power, and economic interests are very easy to see.

Many have wanted to understand challenging official knowledge as an expression of oppositional or alternative social currents that undermine the "hegemonic paradigms" in a society—dominant ways of understanding the world—but this is only half the story. Today, environmental protection and several perspectives that emerged within the environmental movement are in many ways integrated in dominant conceptions of nature. Environmental protection is now an obvious topic on the political agenda, the state apparatus comprises big agencies responsible for it, and environmental science provides premises for governmental action. Even if environmental organizations often act in opposition to government policy, environmental protection is institutionally bound to politics, management, and research. Institutions are populated by people who are all products of the same academic educational machinery and whose conceptions of the world are influenced by the same type of abstract scientific knowledge. These perspectives have become ingrained in a hegemonic paradigm upheld by powerful groups, even by the state in modern societies. It is no wonder, then, that environmentalism is attacked from below.

As we have discussed in previous chapters, expert knowledge is challenged through an active cultivation of lay, practical knowledge with solid roots in social groups who do not belong to, nor feel at home in, the segments that control the hegemonic environmental discourse. Practical lay knowledge (often labeled "common sense") is not only founded in forms of experience other than those molded in a scientific framework but also frequently entails skepticism and mistrust toward science-based insights. Environmental protection (especially in the form of nature conservation), the environmental movement, and environmental government agencies seem to have become favorite targets for some social groups who perceive themselves to be in a subordinate position vis-à-vis those in power. Challenging dominant views in the environmental field is frequently a component in the cultural resistance such social groups may launch to defend their sociocultural autonomy, as discussed in chapter 4.

There is no consensus on large carnivore issues among people who live in areas with large carnivores, as seen in chapter 5. Furthermore, those with similar views—for example, a more or less open skepticism toward official knowledge about the carnivore situation—are not always in the same social position. However, we have found that the concentration of large carnivore skepticism and ill will toward management agencies and employees is strongest among what we call the rural working class: working people in rural areas with strong historical and cultural ties to traditional use of natural resources (see chapter 4). In the wolf areas, we do not generally include farmers and landowners in this category, although they are also typically skeptical of wolves. The positions taken by farmers and landowners may, to a considerable degree, be attributed to adverse economic effects of the wolf's presence. We see the stance taken by propertyless "ordinary people" partly as a component in their resistance against the cultural dominance of the expanding middle class. This is a struggle that plays out in many arenas, certainly not only in controversies over wolves, large carnivores, or conservation and land management. As seen in chapter 3, millionaire forest owners may share many of their views, but rarely all. Importantly, such an alliance between social groups dependent on material production, albeit in very different ways, is not exclusive to the field of environmental politics (Krange and Skogen 2007a; Skogen and Krange 2010; see also Frank 2004).

ENVIRONMENTALISM, SCIENCE, AND THE EXPANSION OF THE MIDDLE CLASS

A clear connection exists between environmentalism and science. While the relation between the environmental movement and dominant science has often been construed as one of conflict (cf. Beck 2000), this is at best an incomplete understanding. Of course, some important incongruities regarding typical perspectives of the environmental movement and those of mainstream science do exist. Most notably, antagonism is apparent between environmentalism and those strands of science manifestly embedded in a utilitarian understanding of nature and, indeed, frequently interwoven with what may legitimately be considered the interests of "industrialism" or "capital." But the environmental movement does not generally appear to be estranged from science as such, and strong historical ties exist between the environmental movement and the life sciences. The predominant environmental discourse today is powerfully informed by science, and environmentalists regularly claim a scientific basis for their arguments, even when this underpins a more fundamental ideological stance. Furthermore, natural science is not monolithic, and much of the information about environmental degradation and risk is conveyed to us by science, although frequently mediated through mass media. Seen from a slightly different angle, the prominence of scientific understanding within environmental discourse is no surprise when we consider the predominant middle-class basis of the environmental movement (Cotgrove and Duff 1980; Kriesi 1989; Skogen 1999; Strandbu and Skogen 2000), as familiarity with the academic field through higher education is characteristic of the middle class. Some of the groups that make up the core constituency of the environmental movement are indeed situated within academia itself, further accentuating this connection. Academics (quite a few of whom seem to have a background in the environmental movement) almost exclusively staff environment management agencies, which helps to generate a discourse largely shared with the environmental movement. Although considerable differences in emphasis may occur regarding basic political issues, as well as in the interpretation of concrete situations, there is generally a common conceptual ground, which provides a familiar and comfortable frame around the exchange of diverging opinions.

The division of labor in modern societies has led to a removal of people from nature, as both producers and consumers, and to a general fragmentation of knowledge that severely affects our understandings of nature. A significant aspect of this division of labor is what Harry Braverman (1974) termed the separation of conception from execution: abstract, scientific knowledge has been separated from lay knowledge accumulated through concrete, everyday experience, and the former has achieved a dominant position vis-à-vis the latter. In traditional crafts, those who planned the production also executed it. In modern industry, a sharp divide exists between those who make plans and those who do the work the plans require. Several classic studies of shop floor culture have vividly described this form of alienation, not least how workers reacted to the fanciful but impractical ideas of engineers and managers (Lysgaard [1961] 1985; Willis 1979).

Abstract scientific knowledge is separated from knowledge developed through concrete everyday experience (Braverman 1974; Dickens 1996; Jovchelovitch 2008; Wynne 1996). But science is itself fragmented. It is divided into disciplines that provide disparate perspectives on "reality," and struggles occur between different tendencies within any discipline and in all fields of research. Therefore, the strength of the connections between science and power will vary. Generally, however, scientific knowledge is in a powerful position in contrast to knowledge based on everyday experience, especially tacit forms of knowledge rarely expressed verbally. Such knowledge may be generated through collective experience and accumulated through generations, but it is not molded into forms recognized by science and lacks the institutional basis required to achieve such recognition. The technological advances integral to the development of modern societies originate from a science that has largely been integrated in capitalist economy. Thus, the centrality of technological development and even attempts at scientific social engineering have reinforced the dominance of abstract, academic knowledge over lay, tacit knowledge.

The emergence and rapid growth of the modern middle class must also be understood in this context. Claims to monopoly over socially useful (abstract) knowledge have been fundamental to (the very successful) middle-class strategies for carving out new, advantageous social positions. An active construction of an incessantly growing need for such knowledge in every conceivable area of human life has been an important force behind the dramatic expansion of the middle class throughout the era of industrial capitalism (Martin 1998; Skogen and Krange 2010). To a considerable extent, the social position of the modern middle class is based on the supremacy of scientific knowledge over lay, practical knowledge. Highly educated middle-class people have a clear interest in maintaining this relationship, one basis for which is found in the multitude of state institutions that have sprouted over the past decades and in other sectors with similar characteristics in terms of knowledge basis, such as the media and professionalized NGOs. Yet even if the growth of the modern middle class is inextricably bound to the development of capitalism throughout the twentieth century, large sections of it are now tied to economic sectors outside of capitalism's core processes, not least in sections of the state apparatus that produce what Erik Olin Wright (1997) terms "decommodified use values." In many advanced countries, these include health services, education, and state-supported arts, but indeed also public wildlife and land management. A core point, then, is that large parts of the middle class are situated outside of the economic engine room and are to a great extent active in areas that appear to be decoupled from a market logic. This separation entails development of cultural traits—interpretation frameworks, value sets—that run counter to the production-oriented paradigm that has dominated modern societies. We see this clearly in conflicts over land use, resource utilization, and conservation. A conservation-oriented view of nature with a distinct middle-class basis is pitted against a view of nature emphasizing resource use typically shared by people who in more direct ways depend on material production.

The alienation of lay knowledge is a consequence of people's knowledge being taken away from them. Lay knowledge is frequently ridiculed and rarely taken seriously by certified "experts." Those who feel they bear the burden of current large carnivore policy are alienated in relation to the knowledge that forms the basis for policy and management. What we see here is a "social division of knowledge." The separation of abstract knowledge from other forms of knowledge, as well as its hegemonic position vis-à-vis these other forms, is of course interwoven with the alienation of labor in capitalism through the social division of labor, described by Karl Marx in the infancy of capitalism. This means that the labor process and its products are taken away from the real producers, in the sense that they have limited or no control over such fundamental processes and are thus estranged from the products of their own labor. In Marx's classical analysis, workers on the assembly line do not act according to their own will but the capitalist's intentions. In a somewhat similar way, "ordinary people" in the wolf areas must adapt to a situation determined by the knowledge and intentions of powerful others. The conflicts over knowledge in the field of large carnivore management and conservation comprise dimensions that are also class conflicts. However, the capitalists do not emerge as the enemy of working people as in the old industrial society, but rather a highly educated and to a large extent state-funded middle class does (Krange and Skogen 2007a; Skogen and Krange 2010).

As explained in chapter 4, people do not sit still and accept domination. In many cases, challenging scientific knowledge about large carnivores may be instances of the cultural resistance we described there, not least concerning rumors about secret and illegal introduction of wolves, which we shall examine more closely in the next chapter. Interestingly, these stories are also examples of how, under certain circumstances, hidden transcripts—alternative explanation models that usually thrive in the background—are brought "on-stage," to the media and to the field of politics, and receive substantial support from powerful groups that see this as serving their own interests.

NOTES

1. Many informants always talked about the county-level environment agency, which in reality is an arm of the national government, as "the County" (Fylket). Not everybody understands the difference between this type of county-level agency and the administrative apparatus of the elected county assembly (which, to complete the confusion, has some overlapping responsibilities, although not for large carnivores). The complexity

of public administration contributes to alienation of many local people and is certainly not unique to Norway.

2. This is not the same as the county-level environmental agencies but rather a sort of ranger service that also has a number of local part-time employees called "carnivore contacts," adding to the complexity (and confusion) characteristic of large carnivore management.

3. It is widely held that one important task for social science is to diagnose the present, that is, to identify the characteristic features of an epoch. Quite a few sociologists are enthusiastic contributors, and our own present has been labeled liquid modernity, post-modernity, late modernity, or high modernity. Authors such as Giddens (1991), Beck (2000), and Bauman (2000) develop slightly different concepts to capture the essence of what is thought to be a new version of modernity, but one common denominator is the lost authority of science.

4. On hegemony, see the introduction and chapter 4.

CHAPTER SEVEN

RUMORS ABOUT THE SECRET REINTRODUCTION OF WOLVES

❄ ❄ ❄

The reappearance of wolves has led to conflicts in rural areas across the globe, as has been thoroughly documented. In Europe and North America the conflict level has risen as wolves have recolonized areas from which they have been absent for decades or even centuries (see Bjerke et al. 1998; Ericsson and Heberlein, 2003; Naughton-Treves et al. 2003; Skogen and Thrane 2008; Wilson 1997). In this chapter, we take a closer look at the situation in the French Alps and compare with our Norwegian study areas.

Some years ago, we conducted a study together with French sociologist Isabelle Mauz from the research institute Cemagref (now Irstea) in Grenoble. We compared our Norwegian interview material to her extensive material from the Savoie area in the Alps. We had already observed that the conflict patterns appeared very similar, and we were particularly interested in the stories about the secret reintroduction of wolves, which flourished in both countries. Our study areas were far apart, but the stories nonetheless seemed almost identical. However, no attempt at analyzing these narratives' roles in the social constructions of the "wolf field" had previously been made. Two varieties of these narratives have become particularly prominent in Norway and France. Stories about shady activities like the secret reintroduction of wolves are common among wolf adversaries. Either extreme environmentalists or an alliance between environmentalists and government agencies allegedly conduct these clandestine operations. Another narrative important to the pro-wolf camp depicts sheep husbandry practices common to Norway and the French Alps (unattended rough grazing) as unique to each region. Norwegian wolf proponents contend that the Norwegian situation is singular, and their French counterparts make the same claim for France. An image of wolf problems as originating from the local farmers' particularly irresponsible attitudes and primitive views of nature exists in both countries. Conflicts with wolves are supposedly almost unknown in other countries.

Although there are similarities among the narratives, we argue there are also significant differences, which should be understood as manifestations of power relations. To this end, we explore the usefulness of two theoretical perspectives: a social theory of rumors, focusing in particular on demonic rumors as a manifestation of cultural resistance, and a theory of symbolic power. By comparing two regions far apart in Europe, we will try to identify social mechanisms of a general nature, mechanisms that are not bound to particular regional contexts. We use interview data from France and Norway, supplemented by written material such as newspaper articles and websites, as the empirical basis for our analysis. We take a "grounded" approach in the sense that we describe the narratives first and then move on to the theoretical frameworks we see as most appropriate for the analysis. This procedure closely resembles the actual research process, where these particular narratives were not the initial focus but where they materialized over time and eventually demanded research attention—and theories—of their own. We feel this approach will familiarize the reader with the subject matter of the study in a way that—hopefully—makes our analytical perspectives come across as logical and well adapted to the data.

THE FRENCH STUDY AREA

The French and Norwegian studies were conducted separately, and data were compared after the conclusion of both projects. However, the studies are well suited to comparison, as similar methods were used. The Norwegian study sites have been presented already, but a brief description of the French project and study site is necessary. The French project started in 1997 as a study of wildlife's role in the symbolic construction of social relations in the Vanoise area. Hunters and national park guards were interviewed for this purpose. However, during fieldwork, things changed dramatically when the wolves arrived. The first attacks on livestock occurred in the fall of 1997. The project was adjusted to focus on the role attributed to wolves in the social construction of nature and was eventually extended to include farmers, conservationists, and various agents of public land management (Mauz 2005). More than a hundred in-depth interviews were conducted from 1997 to 2000. A second phase, which was carried out in 2005 and 2006 and included another twenty-five informants, was directed at local people's reactions to wolf management and wolf population monitoring. Both studies generated the same type of interviews, covering issues relevant to the present analysis.

The study area consists of the twenty-eight small municipalities partly included in the Vanoise National Park. Vanoise is the massif of the Northern Alps and separates Haute-Maurienne (the high valley of the Arc) from Haute-Tarentaise (the high valley of the Isère). Both valleys are close to the Italian border. The interior Alps' favorable climate has allowed extensive agriculture and livestock production to develop, which for centuries constituted the region's economic backbone. Dairy farming, which takes different forms from place to place, is particularly well developed. Sheep farming underwent extensive change in the 1960s and 1970s, when much larger herds raised for meat substituted small herds raised for milk. The importance of sheep farming has fluctuated but remains a main economic activity in some municipalities.

Today, tourism dominates the economy, directly or indirectly providing a majority of inhabitants with all or part of their income. Haute-Tarentaise now has the largest concentration of ski resorts in Europe, the more famous being Val d'Isère, Tignes, Les Arcs, and La Plagne. Although winter tourism is more significant, summer tourism is also important. In the past, Haute-Tarentaise and Haute-Maurienne were characterized by a high rate of temporary and permanent out-migration, but this trend has now halted or even reversed. The population of Haute-Maurienne's twelve municipalities affected by the national park appears to have stabilized, and the sixteen municipalities of Haute-Tarentaise have seen a 78 percent population growth from 13,700 in 1962 to 24,200 in 1999.

HOW DID THE WOLVES GET THERE?

As noted earlier, the expansion of Norway's wolf population started in the late 1980s. Biologists initially believed a few wolves from the native Scandinavian population had survived and multiplied. However, genetic analyses have demonstrated that all of the wolves currently in Scandinavia are of Finnish/Russian extraction and that the native population must be considered lost (Vila et al. 2003). Accounts of wolves moving in from Finland were, until recently, supplemented by explanations of how small wolf populations may undergo rapid growth given favorable conditions. Although this latter mechanism seems to be less relevant now, it was part of the dominant "wolf reappearance paradigm" at the time of our interviews. Throughout 2006–2007, Norway and Sweden shared a trans-boundary wolf population of approximately 170 animals, around forty of which were in Norway more or less permanently, according to official figures. Both the population as a whole and the Norwegian packs were concentrated in a relatively limited area along the southern part of the border.

In 1992, wolves were officially observed, for the first time since their disappearance from France around 1930, in Mercantour National Park of the Southern Alps. The official account states that the wolves migrated from Italy and that such dispersion is a natural process when wolves are not pursued by humans. At the time of our study, approximately 130 wolves were in France. A trans-boundary alpine population shared with Italy and Switzerland amounted to about 170 animals. These explanations of why wolves have come back are based on science and advanced by wildlife biologists and wildlife managers, and thus also by resource management agencies and eventually the political establishment, including the national media. Naturally, such accounts also have the full support of environmental organizations. However, they do not go uncontested. In both France and Norway, alternative accounts of wolf reappearances flourish among those who do not welcome the returning wolves—most notably farmers and hunters with firm roots in a traditional resource extraction culture.

SECRET REINTRODUCTION

Park officials presented the wolves' reappearance in Mercantour as a welcome event, but as soon as the news was publicly known, farmers and hunters offered their own explanation: the wolves could not possibly have returned on their own; they must have been secretly introduced. We encountered the same opinions in our study area of the Northern Alps, where the wolves arrived a few years later. These views were advanced in interviews with farmers, farming organization officials, and hunters. They also appear in material published by farming organizations (e.g. Chambre d'agriculture des Alpes Maritimes 1996). In one farmer's living room was a picture of a jubilant wolf riding a motor scooter with its tongue hanging out, accompanied by the text, "The wolf returns from Italy on a Vespa!" Alongside hard-core wolf adversaries, many farmers who generally subscribe to more moderate views and many hunters who do not take a particularly aggressive stance toward wolves also share this view:

> We are all convinced that the wolves have been released. (...) I know they might come from the top of the mountain but they don't jump like that, don't tell me stories [meaning that they appear in one place and then in some other distant place, as if they had jumped from the first one to the second]. Why didn't they arrive ten years earlier?
>
> (Sheep farmer, Savoie)

In Norway, wolves are also said to have been secretly bred in captivity and released. This version of the reappearance story was encountered in many interviews but can also be found on websites of organizations opposed to carnivore protection and in publications by anti-wolf activists (e.g., Toverud 2001). Furthermore, media coverage of the wolf conflicts, even on national television, has conveyed the conspiracy. As a sheep farmer from Stor-Elvdal put it:

> Yes, I am certain of it: that they descend from wolves that were released. (...) It is a strange thing that the wolves appear exactly where the government wants

wolves. That is some coincidence! They draw a line on a map, and lo and be-
hold, the wolves appear so nicely distributed inside it that you would think
they had used a pair of compasses.

(Sheep farmer, Stor-Elvdal)

There are two purported empirical bases for these stories: alleged observations
of animals being released or fed and of unnatural behavior or physical appearance. Some observations are in themselves not controversial; only their interpretation ties them to wolf introduction.

In Norway, we heard of observations of nonlocal trucks carrying dog cages
on logging roads after dark. And whereas popular lore often ties small aircraft
appearing in remote places after dark to drug trafficking and espionage, in our
study areas they are tied to the secret introduction of wolves. In France, there are
stories about local people who have shot wolves illegally and found microchips
on the animals—clear proof that they had been released. However, because
the hunters have themselves committed crimes, they cannot come forward.
The other type of observation concerns the animals' behavior and appearance.
Whether the behavior in question is in fact something that wild wolves would
not do and whether a particular fur color is outside of the normal range are always open to discussion. Wildlife biologists have ready explanations, but those
who believe in an introduction conspiracy rarely accept them.

The most widespread notion of unnatural behavior in Norway relates to
lack of shyness. Wolves are frequently observed near houses; they have attacked
chained dogs, eaten cat food on people's doorsteps, and lurked around kinder-
gartens in broad daylight. Biologists claim to recognize these behaviors as normal and describe the wolf as a feeding opportunist always on the lookout for an
easy meal (not school children but perhaps the contents of the school's garbage
cans). But most people unaccustomed to wolves think of them as we see them
on television: living in—and presumably preferring—remote wilderness areas.
Compared to this image, urbanite wolves may seem unnatural and frightening.
Not only are they too close for comfort; they may also be unpredictable if they
are raised in captivity and lack the presumed natural shyness:

> And it was strange indeed that it didn't stop in Rakkestad. That's much closer
> to Sweden, and a much larger forest. Suddenly it surfaced here and it wasn't
> scared of anything. Several people had it inside their yards and that was a
> strange thing, don't you think?
>
> (Farmer, Våler)

In Våler, the first wolves had to cross a large river to reach their present location, and they passed through a semi-wilderness that is much more like popular
images of wolf habitats than are the small forest patches where they eventually
settled. People see these actions as clear indications that the wolves did not find
their way to Våler on their own:

Just have a look at the map of Østfold: there you have the [Swedish] border, then Aremark, Rakkestad, Eidsberg, where you have the highest density of moose and roe deer this side of Oslo. Nobody can tell me that the wolves walked through that large buffet and then swam across the big, cold river to get here. I wouldn't question the wolf development if it had started on the other side of the [river] Glomma, and then expanded in our direction. But it started in the wrong end.

(Hunter, Våler)

Biologists retaliate, however, by claiming the density of prey is actually higher in the agricultural landscape. But this argument seems to fall on deaf ears for many reasons, one being that the first litter born here *was* actually hybrid: the alpha female had mated with a domestic dog. Although wildlife authorities eventually culled the hybrids, many stories circulate about intentional breeding and how the general population was never meant to know the cubs were not pure wolves. Rumors also say authorities allowed at least one of the wolf-dogs to escape.

Stories explaining why wolves could not have independently wandered the routes and distances claimed by biologists are also common in France. Genetic analyses indicate that even a wolf that found its way to the Nohèdes National Preserve in the Pyrenees was of Italian origin (*Le Monde,* 28 August 1999), but the great distance it must have traveled reinforces the notions of clandestine reintroduction (*Le Monde,* 8 September 1999, letter to the editor). Wolf opponents also point out that wolves have been introduced in other countries and that, in France, other large carnivores have been officially released (lynx in the Vosges, bears in the Pyrenees):

Why did people accuse us of having reintroduced the wolf? Because there is an image of an administration that reintroduces many animals. So they said, "Why not the wolf? After all, you reintroduce other animals; you won't have us believe you are unable to do that!

(National park guard, Alpes-Maritimes)

Other French accounts also describe unnatural characteristics among wolves. Wolves are said to attack sheep for different purposes: some kill to eat and actually consume their prey, whereas others merely hunt for play and only nibble at the sheep they kill. Because such behavior is not expected of wild wolves, these "small eaters" are thought to be born in captivity:

In nature, when [the wolves] go hunting, they attack an animal and they kill it, normally. Here, they killed four [sheep], they wounded some, and they ate half a kilo! (Farmer's wife, Savoie)

The wolves that arrived at the Glandon pass, those surely came from Mercantour through the Hautes-Alpes, but these ones [that we have got here], they are no real wolves; they are released wolves (...) because they don't attack the same way. They hardly ate anything. And in the droppings [the scientists] found chamois—only chamois [i.e., the wolves killed sheep but *ate* only chamois, a goatlike antelope typical of the Alps].

(Sheep farmer, Savoie)

Some informants also claim the released wolves are a different color than Italian wolves. The same point is made in the Norwegian interviews, as well as in written anti-wolf material (e.g., Toverud 2001). The native Scandinavian wolves, now extinct, were "stone gray," whereas the newcomers are yellowish (supposedly an adaptation to the colors of the Estonian forest floor) or red-brown, allegedly like Russian wolves (Toverud 2001: 76–77).

WHO ARE THE CULPRITS AND WHAT ARE THEIR MOTIVES?

The actors supposedly responsible for reintroductions are not always identified. French wolf opponents often use impersonal expressions such as "one has released them" or "they have been released." When collective culprits are indicated, conservationists are frequently accused, as are government foresters thought to have joined forces with conservationists. Some informants accused a known "wolf lover," a former director of the Directorate for Nature Protection. Conservationists' and foresters' old statements supporting the release of large carnivores to efficiently control ungulate populations, thereby reducing forest damage, have been exhumed. Similar accounts in Norway claim plans made in the 1970s for wolf reintroduction in Sweden, and later officially abandoned, were secretly implemented. Much is made of the fact that the person in charge of the proposed plans now occupies an elevated position in the Swedish Environmental Protection Agency. However, this type of sophisticated reasoning of the actual planning and organization was rare in the interviews and more common in written material distributed by anti-wolf networks (e.g., Toverud 2001) and on the Web. Our Norwegian informants spoke in the same general terms as the French, pointing to "those" who have released wolves. When asked to elaborate, most informants incriminated extremist environmentalists, but resource management agencies and politicians were also frequently mentioned. Some even said a former Minister of the Environment was personally involved.

> I am sure they know a lot more up at Stortinget [Norwegian parliament] than we get to know here. Because I think this is run from the very top. But things like that are almost impossible to prove. Then somebody in those circles would have to blow the whistle. (...) People have seen wolves being released. A lot of people claim that they have seen it. But they don't dare to come forward. There are powerful forces behind it, so people are afraid of their health

if they talk about it publicly. (...) These things have been carefully planned for a long time. Many years. It seems like they have formed alliances with people in high places, maybe even right at the top—people who are pro-wolf, or neutral people who they have persuaded. (...) Because there is so little interest in establishing whether these are pure wolves, I think it is pretty clear. There are obviously strong forces behind it all.

(Hunter, Våler)

In France, many wolf adversaries are convinced wolves had been reintroduced primarily to accelerate the depopulation of the French countryside:

There is nothing we can do. They want to destroy the farmers. The wolf is a means to destroy them. All these politicians, these fat men who earn a lot of money—and we are the puppets. There's nobody left in the countryside. In the village, there will be no more farmers within five years. What is it going to look like?

(Farmer, Hautes-Alpes)

Wolf adversaries consider the wolves "biological weapons" and see themselves as victims of a plot contrived by powerful groups who loathe rural people and their way of life. Norwegian informants and written material in Norway draw the same image, but more modest versions of the story are more common. The harm caused to rural areas is seen not as the chief goal behind the introductions but rather as a side effect of a strategy aimed at reconstructing a scenic wilderness as playground and aesthetic object for city people. In any case, it affects only backward country people with primitive views of nature.

WHO TELLS THE STORIES?

In both countries, these narratives are found primarily among farmers and local hunters, but they also exist in other segments of the population. For example, several big landowners in Norway said "there might be something in it." We discussed the social basis of resistance against wolves earlier. In the Savoie area, farmers and hunters form the core of the wolf resistance. There are no big private landowners there. As in Norway, the material basis for the farmers' engagement is obvious, as the presence of wolves directly affects livestock husbandry. As for the hunters, the situation is different from what we have seen in Norway, although there are also similarities. In the French study area—like in Norway— hunters are predominantly local and virtually all male. One is not considered a proper hunter unless one hunts chamois. Some women and outsiders do hunt, but they are always considered exceptions and several tricks are used to exclude them. The exclusion of outsiders is particularly severe in communities with large ski resorts, such as Val d'Isère, as if hunting chamois remains the only way to prove, despite tourism, the inhabitants still possess and master these places. Put briefly, hunting is a way of stating, "I belong to this place, and I am a true mountain man."

Whereas traditional hunters form a stronghold of wolf resistance in Norway, the picture is different in France. Many hunters are skeptical of wolves, but most appear to be more open-minded than their Norwegian counterparts. Several reasons may account for this attitude, but one striking difference between France and Norway is the fate of the hunters' beloved dogs. In Norway and Sweden, but not in France, wolves have attacked and killed many hunting dogs. Knowing the affectionate relationship between hunters and their dogs and the tremendous amount of time and money many hunters invest in training them, wolves are, unsurprisingly, not popular (see also Naughton-Treves et al. 2003). Indeed, typical Scandinavian hunting methods, which entail the use of untethered dogs, are now seen as impossible in areas with wolves. Because many hunters regard the cooperation with their dog as more rewarding than the actual kill (Krange and Skogen 2007b), the loss of this hunting form is all the more aggravating. So far, for a variety of potential reasons, their French counterparts have not undergone these experiences with wolves. Hunting with dogs seems to be less common and is done differently compared to Scandinavia, but attacks on dogs in France, should they occur, would undoubtedly and significantly raise the temperature in the conflict. Another factor determining the different situations is the historical development of hunting. Hunting as a relatively common leisure pursuit appears to have an even shorter history in France than in Norway, although some forms of hunting now important in Norway are also postwar phenomena (Aagedal and Brottveit 1999). Nevertheless, hunting appears to have firmer roots in rural communities in Norway.

However, in Norway and France alike, a number of economic, cultural, and practical tensions traditionally and currently exist between farmers and hunters. Conflicts of interest regarding access to hunting occur because farmers are often landowners and may want to maximize hunting profits. Conflicts also arise between hunting and sheep: sheep are collected with shepherd dogs and a lot of commotion in the prime hunting season. Hunting dogs occasionally chase sheep and can be shot legally by farmers. (We discussed these tensions more thoroughly in chapter 3.) In France, wild boars, though attractive game to hunters, cause serious crop damage and are seen as vermin by farmers. In Norway, many farmers own forest properties, and young pine trees in particular are eaten by moose, the number one prestigious game species. In Norway the arrival of wolves seems to have subdued these tensions, and we have seen the emergence of a new (probably fragile) alliance among hunters, sheep farmers, and landowners (see chapter 3 and Skogen and Krange 2003). This has not happened in France; on the contrary, the relative lack of open hostility toward wolves among French hunters can be attributed in part to their reluctance to join forces with farmers.

UNIQUE CONFLICTS, LAZY FARMERS, AND OUTDATED ATTITUDES

Conservationists in both Norway and France often claim that modern sheep-herding practices in their region increase depredation problems and that the farmers in their country have particularly primitive attitudes toward the utilization of natural resources and the value of biodiversity. People contend in both France and Norway that sheep farmers in other countries herd their sheep or take other measures to prevent attacks. Therefore, the conflicts between sheepherders and wolves are unique to France say the French conservationists—or unique to Norway say the Norwegians. People in other regions of Europe are said to be astonished to hear about the fierce conflicts in Norway—or in France. The national media seem to have picked up on these stories, and they generally convey the same picture, which seems to have disseminated throughout significant parts of the population in both countries. The sheep farmers rarely contest the uniqueness of their situation but rather choose to defend it as necessary given local conditions and as desirable for the environment and for animal welfare.

Regarding the herding methods, the most important issue is the practice of leaving sheep in the mountains and forests without human supervision except for sporadic inspections. Conservationists claim the sheep are vulnerable to attacks and other accidents when left to themselves in in rough terrain, a practice often attributed to the laziness of modern farmers and to part-time farming. Sheep farming is construed as "easy money" because it is heavily subsidized:

> It's a pity that today 90 percent of farmers of the Southern Alps and even of the Northern Alps are only subsidy hunters: grass subsidy, meat subsidy, subsidy for this, subsidy for that, it must be stopped. Subsidies make up 70 percent of their income. That really is a problem.

(Conservationist, Alpes de Haute-Provence)

Current practices are seen as fundamentally different from the affectionate relationship believed to have previously existed between farmers and their animals, which entailed herding the flocks through the grazing season:

> There used to be many small herds, people lived on farms, and they had many herds but (...) there were always two or three children or the wife who would guard them. Nowadays, it's completely different. It's really an industrial herd, and it is practically never looked after.

(National park guard, Savoie)

In France, current practices are presented as the product of an unfortunate combination of modern agricultural policy and the French's primitive attitudes toward domestic animals and nature. Indeed, wolf supporters changed their argument as the wolves moved north. As long as Mercantour in Southern France was the only affected area, wolf supporters accused farmers from the South of being afflicted by the well-known defects of southern folk: laziness and propensity to exaggerate and cheat. When wolves arrived in the Northern Alps and triggered controversy there, the French in general, and farmers in particular, were said to be characterized by their permanent contesting of the established order, their propensity to disobey laws, and their rejection of nature conservation. The idea is regularly repeated that "all this is very French; everywhere else things are fine, people live with wolves, and there are no problems." Farmers from Eastern Europe (particularly Rumania and Bulgaria) are admired because they allegedly live with much larger carnivore populations but have fewer problems—and if they have problems, they accept the carnivores as "natural" and "valuable" anyway:

> In Rumania, a sheep is much more valuable than it is in France. If the wolf gobbles up a sheep there, it's bankruptcy. Here, well, it's OK, between insurances and compensations and all; they can almost earn more if their sheep are wolfed down than if they lead them to the slaughterhouse. There, they manage to live, although there are approximately 5,000 wolves, in a country which is about the same size as France—5,000 wolves they manage to live with; they manage to live with around (...) 3,000 bears, they manage to live with maybe at least 1,500 lynx, without this raising problems, only because shepherds do their job. There are predators so there are problems, so they organize themselves accordingly. There are guard dogs, there are people staying all the time with their herds in the mountain, leaving the herd unattended when dusk comes is unthinkable, herds are gathered next to the shepherd's shelter, animals are not abandoned, and it works.
>
> (Conservationist, Alpes de Haute-Provence)

Their Norwegian counterparts paint a similar picture; in fact, the interview material contains statements on the idyllic situation in Eastern Europe that could have been translated directly from French:

> In Rumania, there are 2,500 wolves, and probably more sheep than here. But there they have hired shepherds to look after the herds. I think they could do that here too, instead of shooting the poor wolves. I think the wolves should be left in peace.
>
> (High school student, Trysil)

The Norwegian discourse is dominated by images of current herding practices as something that could happen only in Norway because sheep farmers receive such generous subsidies and because Norwegians are accustomed to using nature as they please for their own benefit but have lost touch with the traditional ways of sustainable resource utilization:

> [The sheep] is not an animal that is adapted to a life in Norwegian nature. No matter how much that statement provokes farmers, it is a fact! (...) Except for reindeer, we don't have domesticated animals that are adapted to a life on their own in the backcountry, and we should act on that knowledge. If we are going to produce meat, then we must use animals that are adapted [e.g., older sheep breeds] and the others we must take care of in other ways. (...) People who have assumed the responsibility of owning an animal also have an obligation to know the fate of that animal. And to do everything that is possible to prevent it from suffering. (...) I cannot accept a form of husbandry where animals are sent on their own into an environment that they are not adapted to, with the result that close to 150,000 animals die, often in great pain, during the four-month grazing season. I don't think that is ethically acceptable.
>
> (Conservationist, Stor-Elvdal)

Along with farmers from other countries, farmers from the past are used as positive examples. The old-time farmers are supposed to have had much closer relationships with their animals. Their active shepherding allegedly prevented carnivore attacks so that the relationship between farmers and large carnivores was much less strained than it is today—much like the somewhat mythical situation in contemporary Rumania:

> People knew that the wolf is an animal that is very easily scared. It killed livestock, OK, but it did not kill more animals than the number of natural deaths in a herd.
>
> (Conservationist, Isère)

However, the elderly people interviewed for the French study shook this idyllic picture. They said sheep were not always looked after, there were no guard dogs, and large carnivores were as unpopular in earlier times as they are today. We have no data on the historical relationship between farmers and predators in the Norwegian study areas. However, historical literature generates a similar impression: things were not as idyllic as the conservationists would have people believe. For example, historical documents indicate that fear of wolves was common (see Snerte 2001) and even that wolves were frequently accused of killing people (which they have indeed done, the last case being in 1800) (Linnell and Bjerke 2002).

Farmers that are distant in space *or* time are held as models for modern sheepherders. This approach has the obvious advantage of avoiding too close scrutiny, particularly if those who are targeted generally have limited access to information that *could* have thrown some light on these rather simplistic notions of harmony. Such counter-evidence does exist. For example, a French study concluded that current and historic wolf damage to livestock in different countries varies according to herding practices and hunting pressure and that the situation in the Alps is in no way unique (Garde 1998). But this is generally not known to the people who would need the evidence to build their argument. We will return to why in the final section of the chapter, as it can tell us something about the power relations at play in this field.

RUMOR AS CULTURAL RESISTANCE

The wolves' origin has also interested social scientists, at least in France. They have focused on the reintroduction narrative, as well as its supporters, and have noticed it strongly resembles other phenomena studied in the past few decades (Campion-Vincent 2004, 2005b). This is not the first time people have claimed that undesirable species were released—accidentally or deliberately. Many New Yorkers, for instance, are convinced alligators live in the city's sewers (Campion-Vincent 2000; Kapferer 1990). In several French regions, a widespread conviction claims helicopters drop boxes containing vipers and foxes (Campion-Vincent 1990). Some informants explicitly established the connection and said wolves are being released "exactly like vipers."

Such stories are often labeled rumors, commonly considered a derogatory term equaling "gossip." However, the term may also be given a useful scientific meaning. We first introduce a definition developed by Jean-Noël Kapferer (1990): a rumor is the emergence and circulation of a collective interpretation of a problematic event that official sources deny or have not yet confirmed. People tend to repeat a rumor, to contribute to its transmission, or even to nourish it because they are seduced by its content, particularly because preexisting opinions and interpretations are reinforced. Rumors are not necessarily false, but they are unverified. Rumors are counter-narratives, providing alternative explanations less open to scrutiny than the official story while being more exciting and disturbing. Rumors are rarely simple if they can be complicated. Obvious interpretations are often rejected and replaced by more convoluted reasoning. Rumors are often "black," according to Kapferer, in the sense that they present a negative interpretation of events considered problematic; they tend to attribute what actually or fictively happened to persons or collective agents in such a way that they are discredited or dishonored. Rumors are flexible; they spread rapidly and are likely to turn objections and denials from authorities into new arguments in support of the interpretations represented by the rumors (Campion-Vincent 2005a).

The wolf reintroduction narrative is clearly in opposition to the official account. By incriminating state services and scientific institutions, wolf opponents launch resistance against the power of the state and its associates, the biologists, and urban conservationists. The narrative is indeed "black," as it denounces the scandalous existence of a wolf reintroduction network or even a secret alliance comprising people in high places. Furthermore, objections and arguments supporting the official version often do nothing but strengthen the narrative. Hence, wolf opponents' reintroduction stories possess all of the characteristics of rumors outlined by Kapferer. They are among those particularly stubborn rumors that are almost impossible to refute. It is difficult, perhaps impossible, to prove that something did *not* happen (Fine and Turner 2001; Kapferer 1990). How could one show, for example, that no wolf has been intentionally or accidentally released? The question, therefore, in a strictly logical sense, remains open.

Rosemary Coombe (1997) introduces the concept of demonic rumors. The term "demonic" does not mean such rumors should be understood as comprising elements of the supernatural but rather that they establish the existence of malevolent forces responsible for everything from mild social tensions to economic decline and the spreading of AIDS. Coombe focuses on rumors that target multinational corporations like Philip Morris and Procter & Gamble, as well as more modest brand-name companies. In late modernity, capital is "disembedded" and tied to neither place nor personal actors in recognizable ways. Brand names are omnipresent and play a more important part in people's lives than ever before. But the actual production is unseen—its location hard to discover—and, in any case, probably on the other side of the globe. No personal actors seem to be associated with the products, and if such actors are known, they, too, are usually distant and inaccessible. At the same time, the ever-increasing presence of brand-name products in people's lives clearly shows that those who produce them exert tremendous power. Yet there seems to be no way to confront this power, as there is never anybody to confront.

Rumors that incriminate such operations flourish predominantly among people with limited access to traditional political power and who stand to lose the most as a consequence of current processes of economic and social change (Fine and Turner 2001). Coombe contends that stories about malevolent intentions of huge corporations—in particular, efforts to hurt certain disadvantaged ethnic or social groups through poisoning, spreading of disease, or associations with the Ku Klux Klan—efficiently serves two purposes. First, they connect power to agency by introducing purpose and planning, thus making sense of an otherwise vaguely felt association between strenuous social conditions and the omnipresent yet unapproachable economic conglomerates. At the same time, central characteristics of rumors (difficult to trace, even more difficult to come to grips with, always developing in new directions, and deftly absorbing counter-evidence) are similar to how the presence of these conglomerates is felt in people's lives. Thus, demonic rumors could be seen as a way of turning the "weapons" of corporations against them, as corporations appear to be almost as defenseless against rumors as ordinary people are against the economic presence of corporations. Rumors have actually forced giants like Procter & Gamble to use enormous resources to counter them, without success, and smaller companies have been forced out of business (Coombe 1997).

Coombe focuses on the economic forces of modern capitalism and how demonic rumors constitute one way of relating to them, which may fill important functions for people who otherwise feel powerless. This understanding can extend to the way people grapple with the modern state or, indeed, in our case, perceived alliances between the state and the environmental movement. The state exerts power in ways many see as incomprehensible and arbitrary or, worse, as part of a strategy to depopulate rural areas. To some groups, particularly people rooted in traditional forms of resource utilization and consumptive outdoor recreation, modern policies of nature conservation are prime examples. In summary, rumors can be seen as narratives that interpret and explain disturbing aspects of the world otherwise perceived as incomprehensible, diffuse, or explained in unsatisfactory ways. Veronique Campion-Vincent (2005a) claims they constitute a "folk social science." To the extent that they challenge powerful groups' hegemonic paradigms and interpretations, they may also be seen as forms of resistance (Samper 2002).

Stories about wolf reintroduction are quite often elaborate and include chains of reasoning that cannot always be dismissed outright. Some are based at least partially on real observations and extensive knowledge of the areas in question. Nevertheless, wolf supports ridicule them as folklore or preposterous fabrications. People who subscribe to these stories are rejected as either dimwits or conspiracy theorists with hidden agendas. We now turn to a brief discussion of the narratives that thrive in the pro-wolf camp and try to discern why these enjoy a different position.

THE NATURAL RECOVERY THEORY AND THE NOTION OF NATIONAL UNIQUENESS: PRO-WOLF NARRATIVES WITH A STRONG MESSAGE

We say narratives in plural because if we consider the natural recovery theory as a background of "solid scientific fact," against which the oppositional introduction rumors are played out, we would be overlooking its function as a narrative and as an interpretation. The natural recovery theory is a narrative that carries significant cultural meaning. First, it cannot be verified in a strong sense. Even if reintroduction is considered unnecessary and controversial, definite proof that it has never happened does not and cannot exist. Thus, a rejection of reintroduction stories is no less value-based or normative than their support: both are based on a valuation of the credibility and status of the storyteller.

Contrary to the reintroduction rumor, which incorporates new elements as wolves colonize new areas and as allegations are met with "facts," the official explanation itself does not change appreciably. Expansion of populations is a normal phenomenon that need not be elaborated. It is not "black": it blames nobody and reveals no conspiracy. The natural recovery theory does not qualify as a rumor as previously defined, but it is certainly what Campion-Vincent (1976) terms an "exemplary story." For centuries, people have fought wolves with every available device and total destruction as the goal. Yet wolves have been stronger in the long run:

> It's an animal that has always been persecuted, and finally it has stood up against everything and finally it comes back with force (...) and this is a strong image for me. No matter how hard we tried to destroy it, there it is: it is back. And I think it will always come back, whatever we do.
>
> (Conservationist, Isère).

The wolves are powerful symbols of wilderness, and if they can recover, wild nature may not be doomed after all—even in our dismal times.

Supporters of the official version, like those of the reintroduction rumors, subject it to little scrutiny. Although there is no proof that wolves have been released, neither is there proof of their spontaneous return. The explanation is accepted and repeated, not because its veracity is certain but because it is seductive: like its rival, it is a satisfying explanation, reinforcing preconceived notions of the wolf and of its human adversaries. The natural recovery theory and the reintroduction rumors differ in several respects, but they also share some features. Both are desirable explanations that their followers do not want to question and that may function as exemplary stories: one a story about a malevolent conspiracy against rural interests and the other a success story where a former loser is rehabilitated and returns as "the victorious victim."

The theory of natural recovery and the image of national uniqueness appear to share a hegemonic position unlike that of of the introduction rumors. We will extend our discussion of the function of the national uniqueness narrative later. Suffice it to say here that it too is a narrative with a powerful message. It is definitely not a description of "reality" but rather based on a valuation of certain husbandry practices in relation to a desired environmental state—indeed, of farmers as a social group. Like the natural recovery theory, the narrative is not a rumor in the sense we have used the term. It is, however, a strong value statement but one that does not need to make use of the characteristic features of rumors and that originates outside of social segments where oral traditions are still important, albeit in rudimentary forms.

The narrative of national uniqueness contains demonstrably false components yet is met with alternative accounts to a very limited degree, and is widely accepted to the extent that sheep farmers also propagate its central aspects. The narrative of wolf reintroduction, which cannot logically be demonstrated to be completely false, is ridiculed and overrun by a dominant, official story but one as value-laden as the introduction conspiracy narrative itself. The introduction rumors can be seen as cultural resistance against this dominant narrative and the power structures that sustain it, but the struggle is an uneven one: research reports, government white papers, and national media are pitted against oral history and homemade webpages. Why is the situation so unbalanced? The answer must have something to do with power.

SYMBOLIC POWER

The concept of "symbolic power," introduced by Pierre Bourdieu (see Bourdieu and Thompson 1991), seems to hold some promise here. Power is exercised in many ways, some of which are extremely subtle. Indeed, says Bourdieu (Bourdieu and Thompson 1991: 164):

> We have to be able to discover [power] in places where it is least visible, where it is most completely misrecognized—and thus, in fact, recognized. For symbolic power is that invisible power which can be exercised only with the complicity of those who do not want to know that they are subject to it or even that they themselves exercise it.

And further (196): "As instruments of knowledge and communication, 'symbolic structures' can exercise a structuring power only because they themselves are structured. Symbolic power is a power of constructing reality" and in such a way that even the dominated take it for granted. Thus, symbolic production is an instrument of domination but not in the sense that it mechanically reproduces and reinforces capitalism's economic power structures, although there is a strong link here. The field of symbolic production enjoys a relative autonomy, and a struggle occurs within "the dominating classes" over the "hierarchy of the principles of hierarchization" (168). Different groups of specialists have always played a lead role in symbolic production. In our present epoch, this means the new middle class is in an influential position. The symbolic producers will emphasize the superiority of their own specific assets (or "capital," to stick to Bourdieu's terms) within the field of symbolic production. Today, this will include some perspectives on nature that depart from the utilitarian ones that still enjoy a strong position in the current phase of modernity.

Studies conducted in several countries during the past forty years have concluded that the environmental movement derives its fundamental support from middle-class groups with higher education, employment in "nonproductive" sectors, and incomes in the medium range (Cotgrove and Duff 1980; Kriesi 1989; Skogen 1999; Strandbu and Krange 2003). A plausible interpretation of this support is that the exploitation of nature is an integral part of the global process of modernization and rationalization. This opens up new fields of conflict (e.g., over the environment), not least because growing parts of the population, particularly the new middle class, experience relative independence from capitalism's material production process and related core processes (e.g., see Eder 1993; Wright 1997). These social segments' relatively limited influence on important economic factors leads to an increased emphasis on alternative values (Eder 1993; Skogen and Krange 2010). Whether they actually pursue an anti-materialist lifestyle is another matter, but it has become an ideological beacon for many of them. These perspectives, however, also tend to include the notion that all forms of resource extraction are essentially detrimental and that nature should be protected against human activities as far as possible. In this context, conservation of all species, including large carnivores, is an absolute imperative.

Bourdieu may be criticized for drawing an overly deterministic picture of power relations, through not only the theory of symbolic power but also other core concepts, like that of habitus. All of this internalization of power relations and dominant worldviews (of the powerful) seem to leave little room for change. Regardless of whether Bourdieu can be read this way, the concept of symbolic power obviously cannot exclude the concept of resistance, for the acceptance of dominant worldviews is not complete. Indeed, that various counter-interpretations thrive in the background is probably more typical. James C. Scott (1990) claims subordinate groups create a secret discourse that represents a critique of power spoken behind the backs of the powerful—what he terms "hidden transcripts." When the situation calls for it and when conditions allow, the hidden transcripts are brought on stage on-stage and spoken directly in the face of power. Such is the case with the reintroduction conspiracy narratives: they may flourish first and foremost among farmers and within the rural working class, but they are also promoted actively in open defiance of the official and dominant accounts.

What is the social basis of symbolic power in nature management and, more specifically, wolf protection? Narratives of natural recovery and national uniqueness originate within a conglomerate of biological science, resource management agencies, and the environmental movement. The middle-class basis and corresponding cultural profile of the environmental movement means that these narratives originate among what Bourdieu and John B. Thompson (1991) call specialists in symbolic production, who also share central discourses with science and the state. They thrive primarily within class fractions that master and even mold hegemonic cultural forms and are inextricably bound to higher education and academic knowledge. This is a powerful source of domination and thus of conflict (Dickens 1996; Dunk 1994; Skogen 2001).

Why is it important for wolf supporters to stress that farmers are both primitive and isolated? To suggest a strategy of "divide and conquer" is tempting, but we have no basis for claiming such strategic thinking lies behind this narrative's development. Nevertheless, the claim obviously serves a purpose: if local farmers are different from farmers elsewhere, and if they are inordinately incompetent and careless, then they must obviously change their ways and nothing is wrong with the wolf. However, this chain of reasoning is repeated ceaselessly without much scrutiny. Those who present it are rarely willing or capable of investigating its empirical basis. Many rank-and-file conservationists probably believe as fact that French or Norwegian sheep farming *is* unique, as they have no alternative information. But within the environmental movement at large, with its international network and efficient information flows, *some* awareness of the larger patterns must exist. Therefore, to assume a certain degree of intention in the construction of the national uniqueness image does not seem entirely unjustified.

And here is where symbolic power enters the picture: sheep farmers seldom question the assumption that they are different, that their herding practices are unique. In fact, in the Norwegian material, some of the strongest accounts of national uniqueness came from people associated with farming rather than from wolf supporters. Sheep farmers had accepted their singularity as a truism, which is an important dimension of symbolic power: the subalterns *accept* the dominant view;

> A lot of things could be said of our grazing practice, but I think it is close to perfect. There are lots of people who think it is horrible—only in Norway do we have this strange and unreasonable form of sheep husbandry—but I say we should be damn glad of that, that we don't do it in any other way.
>
> (Sheep farmer, Stor-Elvdal)

The negative image of farmers does not have an official status. The state economically supports sheep farming, compensates losses, supports preventive measures, and attempts to accommodate the interests of sheep farmers when wolf protection policy is being shaped. However, the image of rural backwardness and lazy farmers with primitive attitudes may still enjoy a strong position. Indeed, it may demonstrate the somewhat schizophrenic attitude of the state regarding rural development—protecting, to some extent, the interests of the farmers on one hand and implementing numerous measures to "modernize" and adapt agriculture to a globalized economy on the other. As a consequence, many small farms in marginal areas typical of the French Alps and large parts of Norway are forced to develop new sources of income—something the authorities strongly encourage. Thus, the image of farmers who insist on sheepherding in a productionist framework as anti-modern and stubborn is in accordance with official policies concerning rural areas. In this sense the picture matches the type of rural resident the state no longer wants: one who clings to an inflexible and outdated "mode of production," blocking creative adaptation to a new era. This image is also well adapted to a typical new middle class position on conservation—and thus on large carnivore protection—based on a view of nature in which human activities are generally harmful. From this perspective, downscaling agriculture and traditional resource extraction could indeed be seen as contributing to conservation.

So why don't alternative explanations counter the national uniqueness image in the same way as the natural recovery theory? At least two factors could reasonably play a part. First, seeking support and documentation in distant places is not a strategy near at hand for people closely tied to a particular locality, who may have such local attachment as a core element in their identity projects (see chapter 4). The reintroduction stories, while sometimes expanding into the realms of national politics, are always about activities that allegedly take place precisely in the areas where these people live. To the extent that the rumors are based on observation and "data," they are accumulated locally or recounted by local people. Seeking information about conditions elsewhere in the world to counter images drawn by people known to be internationally well connected would require quite different data collection techniques and a suspicion the dominant narrative is flawed. The type of literary liberties obviously permitted in the reintroduction conspiracy stories could probably not be tolerated, as postulates about husbandry practices in other parts of the world could actually be checked quite easily (that "the other side" dares to discount this possibility says something about the power relations at play here).

Second, the image of local uniqueness is not necessarily unpleasant to people who identify strongly with a particular place and may deliberately construct their identities in opposition to current social forces of urbanization and globalization (Krange and Skogen 2011). Also, the argument may be turned around in yet another way: by denigrating livestock production elsewhere, describing it as industrialized, inhumane, unsustainable, and unable to provide healthy food. Local practices are seen as the opposite of all this and thus should serve as an example to governments and farmers in other countries:

> The mortality rate is higher in other countries, if they graze in fenced-in pastures. If you are going to do that, then you get parasite problems. Then you have to vaccinate, and they do that in other countries. In Australia and New Zealand and the USA there is anti-parasite treatment of the wool and the meat all the time; they use antibiotics and medicines of all kinds all the time, and we don't do that so much (...) So, on the whole (...) I see it as a very sensible and rational way to do sheep farming.
>
> (Sheep farmer, Stor-Elvdal)

Our analysis shows that wolf reintroduction rumors may be seen as "folk social science" (Campion-Vincent 2005a) or "hidden transcripts" (Scott 1990) in that they help people make sense of a troubled situation where power structures are difficult to grasp and seem impossible to confront. The rumors may be taken one step further, to open defiance, and serve as means of cultural resistance, actively challenging the dominant wolf recovery paradigm. In this capacity, the rumors are part of a well-stocked cultural resistance toolbox, whereby some subordinate groups challenge social trends perceived as economically and culturally threatening. As we saw in chapter 4, these tools may be effective in bolstering a sense of autonomy and self-esteem, but they rarely have much effect outside of the cultural realm. They cannot provide the political clout required to change development trends in rural or large carnivore management in any significant manner. They may, however, be absorbed and utilized by actors with more influence; actors that have converging interests precisely in the large carnivore issue—interests best served by broad alliances.. The symbolic power the resistance rumors are up against effectively reinforces the hegemony of the official, science-based version of the wolf story and thus also contributes to a solid fundament for the current management regime, which is indeed based on a perspective on human relations with nature that is different from a traditional rural view. The alternative accounts are relegated to a realm of popular lore and conspiracy theory, whereas the notion of nationally unique husbandry practices gains a semiofficial status and the natural recovery theory is typically seen as undisputable scientific truth.

NOTE

This chapter is a revised version of Skogen et al. 2008.

CHAPTER EIGHT

MANAGEMENT OF LARGE CARNIVORES
OPINIONS AND RESPONSES

❄ ❄ ❄

We turn here to people's firsthand experiences of large carnivore management. The issues we discuss are largely determined by those our informants accentuated. Although the use of an interview guide ensured the airing of what we perceived as relevant issues, the nature of the qualitative interview ensures that what most concerns people receives the most attention. Generally, this is an advantage. Discovering what concerns people—and what they are less interested in—is scientifically important and of interest to policy makers and managers. Of the themes we brought up, some caught our informants' attention and led to discussions and vivid accounts. Other topics, including some that are, in principle, important for the management of large carnivores, barely elicited a response.

This latter phenomenon we call a negative finding, an example of which has been mentioned already. Interviewees showed little interest in talking about details in the formal structure of large carnivore management or the decision-making procedures that determine the current regime. Whatever opinions they might have about carnivores, our informants were much more concerned about the general political level (to which they often referred in fairly vague terms but with keen interest) and highly specific local issues. The regional level, which includes among other things the regional management boards, was at best favored with general comments. A few knew the names of one or two members of "their" board, and some knew how board members are appointed. But in general, interviewees steered the conversations over to other issues they obviously felt mattered more. Therefore, we will not discuss perceptions of the regional administration system as a separate topic. Nevertheless, we address this negative finding in more detail later.

This chapter cannot be confined to wolves. Most management arrange-ments, except the designated wolf zone, are essentially the same for all large car-nivores, and to single out a particular species in our talks about management was often impossible and even pointless. People are more used to bears and lynx than wolves in many places. Although there is often great popular interest in wolves (as we have described), people usually know more about the practical management tools devised for other species, and in interviews, they tended to talk about their experience of how these mechanisms work in practice. These opinions on large carnivore management as informants know it firsthand is the topic for this chapter, which is quite descriptive since we want to let people's experiences speak for themselves. There are more quotations in this chapter, and our interpretations and conclusions are modest. We believe this approach should give the reader a more immediate feel for the interviewees' own expe-riences of their reality. We also anticipate that readers who have followed the conflicts over large carnivore management in countries other than Norway will recognize many of the controversies we (or rather, our informants) recount here.

RESOURCE USE IN MANAGEMENT AND RESEARCH

A widespread perception exists among our informants that the management of large carnivores is too costly—an opinion shared across the spectrum of atti-tudes toward large carnivores. Even some who are very much in favor of large carnivores, including wolves, also express this view. Monitoring is one activity often said to cost far too much, which includes tracking, radio tagging (currently, GPS collaring), and, increasingly, camera traps, although these techniques are used as much for scientific purposes as for management. Though there are ex-ceptions, tagging and camera traps are not normally used for management pur-poses, which shows how difficult it can be for people to distinguish scientific research from management. However, not many seem particularly interested in the finer details here anyway. Scientists are often seen as belonging to an alliance that protects large carnivores, and science and management are often treated as one and the same, not least when it comes to population monitoring and movement tracking. People commonly believe monitoring and research both draw resources from other areas, for example, the management of other wildlife and research on other species, as well as completely different policy areas, such as health care and roads.

Resource allocation is often seen in conjunction with other aspects of mon-itoring and research, mainly the ethics of collaring, handling, and chasing an-imals, about which many are critical (we return to this subject later). Many of our informants, regardless of their opinions on the government's large carnivore policy, expressed strong feelings on these issues. An enormous amount of resources is invested in an activity that many find both unethical and unnecessary. Wildlife management, some maintain, makes nature—not only the large carnivores—less natural: nature is denatured or domesticated. According to a participant in a focus group of neighbors from a small hamlet near Halden:

> In principle I'm opposed to ... I can almost say I'm against the management of all wild animals, including large carnivores obviously. I'm against managing them because we humans, we have to control everything. Why can't nature manage itself? And if there's a problem somewhere, like too many predators and they kill too many sheep, for example, why can't people just sit down and talk it over and solve the problem there and then, instead of spending millions on management, which is a load of rubbish in my opinion anyway?
>
> (Neighbors group 1, Halden)

As previously noted, there is a widespread impression that scientists and management officials understate the size of carnivore populations. If you believe populations are growing all the time, spending so much on monitoring might seem particularly unnecessary, as illustrated by a focus group conversation with people from the tourism business in Trysil:

> A: I would agree with people who think the management [system] is on the wrong track. There's too much—it's become—what shall I say? They hold their cards too close to their chest, the scientists, about the real size of our populations of large carnivores. I'm one of those who think the numbers are far higher than reported. There's another thing: they exceed the regional population goal that has been set, and the national population goal that has been set, and which parliament has determined actually. There's quite a lot of agreement on this [population goal] I believe, but when you exceed it, that's plain wrong. And then it's easy to take things into your own hands.
>
> Interviewer: There's one thing I've been wondering: whether you (...) believe the scientists and people at the County Governor's office sort of know there are more but say there aren't that many. It's one thing if they're mistaken—
>
> [Diffuse laughter around the table, humming]
>
> A: Well, I believe they're mistaken [Several others: Darned right!] to a certain extent. You can't say any more than what you ... as a scientist—well, you have to be absolutely certain. (...) You can't go around "believing"! [Several others: No, you can't!] It's easier for someone like me to say there are probably thirty bears slinking around in these parts in the course of year. That's not many, but that's not the point, but they sure get themselves noticed where they roam.
>
> B: It's very risky to say that scientists withhold information; I think we need to tread carefully here.
>
> A: But that's what people think they're doing, and that's a problem. (...) In any case, it's an indisputable fact, that there are [more now]. I'm often outdoors in connection with my job (...) and I'm always seeing tracks and carcasses.

Maybe you get better at noticing them, I won't deny that, like your eyes learn what to look out for, so you spot things you wouldn't have before but that's not the whole reason, that's for sure. This is just my opinion. But when you're outdoors, and you're always seeing tracks like I said, the example I described at the start, when you actually see signs left by three lynx, a wolverine, and two wolves on the same afternoon in a limited area, there's a [good number] of these animals [around].

(Tourism group, Trysil)

In the excerpt above, we hear people who are receptive to the presence of large carnivores in their neighborhood, as long as the populations are kept at what they see as a reasonable level. Here is a hunter with a more critical view of large carnivores but who ends up drawing a similar conclusion:

A: [It's easy] just being outdoors in the forest, having it as a job, and not having to show what you've done. Any other poor fool, like a carpenter, he has to show how much he has done in a day, but the [biologist], he doesn't have to show anything, and what he does show for it is dead wrong anyway. In any another company, they would have thrown him out years ago.

Interviewer: Yes, but I guess they would have given a slightly different account of the situation..

A: Yes, yes, fair enough. They may well believe they're doing ... what they reckon and believe is correct, but—now I don't want be so pig-headed that I can't put it differently, but the job they're doing, at the end of the day, is plain crazy. And then they're believed and listened to. But they [also] get many questions [which they don't answer], "How much did this cost?," "Is there a use for it?" (...) Like for example they've radio tagged young male bears, they've been doing it now—how many years is it, twenty? Now I ask you, what's the point, what is there left to find out? So, they are protecting their own jobs. So we've got to tag more [animals], because we need a justification here and a justification there. It's obvious; they're terrified of losing their jobs, those folks. They certainly haven't managed to find out how many bears there are, that's for sure.

(Hunters group 1, Trysil)

The reason people criticize the use of resources appears to be quite common among all who are concerned about it. First, the money should have been spent on other things (whether nursing home beds or research on other species of wildlife). Second, scientists and government officials live off taxpayers' money without doing proper work. But depending on their stance in the carnivore debate, people also have different explanations as to why the use of resources is unnecessary. For instance, many believe money is unnecessarily spent on preventive measures such as erecting fences. Those who believe there are too many large carnivores may argue that smaller populations would reduce the need for costly and impractical preventive measures. People positively inclined toward large carnivores tend to require farmers to take more responsibility or reduce rough grazing in areas where there is a risk of attacks. So although the reasons for criticizing the management agencies' use of resources might differ to some extent, their criticism comes across the same. People with less interest in the matter think it all costs too much anyway:

> A: But there's another thing: it costs a formidable amount. Has anyone figured it out, how many millions it costs to keep the wolf—can we afford it? It's an interesting question, because we're talking about colossal sums ... just in compensation, you know.
>
> Interviewer: Indeed, compensation and prevention too, fences and—
>
> A: Could have spent the money on something that made more sense, you know. If it costs 100 million, what couldn't you have got instead—
>
> B: [Instead] of the wolf? Nursing home beds. [Laughter]
>
> (Neighbors group 1, Halden)

THE WOLF MANAGEMENT ZONE AND THE IDEA OF ZONE MANAGEMENT

As explained in chapter 1, Norway is divided into eight carnivore management regions with politically appointed boards. The wolf is the responsibility of the two boards that share the designated wolf zone. Within this zone, a politically determined population goal applies. However, the wolf zone represents a management strategy with little support among our informants. One concern was the very idea of dividing habitats for wild animals into zones:

> A: Well, I think it's plain ridiculous. If you ask me, the wolf needs to be where it naturally belongs. Humans can't just draw a line and decide that's where the wolf can live. A wolf can't see a border, and when the regional board gets it into its head to suggest that in the areas where there are extra problems with wolves, then they go and recommend rounding up a pack and moving it someplace else to somehow distribute the burden. Makes you wonder—makes you lose confidence in the boards altogether.
>
> B: There should've been a committee that rounded us [humans] up, you know, and spread us around. That's what it's all about. If there's space, why can't [the wolves] stay here? That's what I'm always asking myself.
>
> (Neighbors group 2, Halden)

Informant A, who lives in an area where wolves are regularly observed, clearly feels it is wrong for people to manage the wolf at all by drawing up boundaries. During the interview, he mentioned several times his great love of nature and its diversity, of which the large carnivores are a part. An inherent contradiction exists between pristine nature and active management, he believes. We have come across this idea on several occasions, maybe especially when discussing collars and GPS transmitters, and the more practical point he touches on is widespread: the wolf sees no borders, and creating zones is acting contrary to the wolf's nature. Many share this opinion, regardless of their position on the question of wolves in Norway.

Another assessment of the zone concerns its size:

Interviewer: What do you think about this type of regulation [like the wolf management zone]?

A: It's a bit constricted.

Interviewer: Constricted?

A: It could have gone further west and north—Engerdal [municipality to the north of the zone]. I reckon they've got off easy up there.

Interviewer: So you think the zone is too small?

A: Yes. Too narrow along the Swedish border. (...) It's not easy getting any fully Norwegian [packs].

(Conservationists, Halden)

The main point here is not the idea of a wolf management zone, per se, but rather its size. We can mention two factors. Again, we encounter the sense of skepticism concerning poor alignment of the wolf's nature and management actions. The zone is small, and if we want wholly Norwegian wolf packs, then more space is needed.[1] Another criticism here: the size of the zone and the population goals set by parliament are incompatible. These conservationists are implicitly berating the authorities for choosing the wrong tools to reach their own targets.

One alternative to a wolf zone is obviously no zone at all:

B: People should speak for themselves on this one, I think. [General laughter.] But I'd put it like this: the [people of Trysil] have managed the large carnivores round here for nearly a thousand years, and they totally exterminated the wolf, (...) [But they] could live with the other species. It was like that between 1873 and 1978, I guess, and then the wolf came back. The old farmers found the necessary balance.

Interviewer: And it would have resulted in the same thing throughout the country, then, if they'd only—

B: Because it would be evenly distributed across the country. Because if they were disturbed more the carnivores would disperse much more than they do today—if they were persecuted. So they would scatter much quicker than now, when they are allowed to empty area after area of wildlife and then move bit by bit.

Interviewer: But the wolf can't—

B: The wolf would disappear of its own accord if that [arrangement] was put in place.

(Neighbors group, Trysil)

These informants do not enjoy wolves as neighbors. They emphasize that wolves cause problems for people who live inside the wolf zone, while those who live outside get off lightly, making the zone unfair. This line of reasoning is the opposite of what we heard from the neighbors group in Halden, but the conclusions are the same. The world would have been a better place without the wolf zone. The difference is that some are concerned for the wolf's welfare, others for the people who live and work within the zone, whom they see as negatively affected.

These quotations show the variation in people's perspectives and arguments when it comes to the wolf zone. Important differences relate to how the individual informant is affected by carnivores' presence and which side they take in the carnivore conflict. What is more remarkable, however, is that they all agree the wolf management zone is a bad idea. The excerpts above are only a small sample of the statements reflecting opinions we encountered many times during our research, but they illustrate some basic features of the zone's widespread criticism.

Inconsistency between political objectives and concrete management efforts is similarly vilified. Quips about wolves' inability to see borders were made, which is part of a general ridicule of authorities that prevails on both sides of the conflict. This brings us to the factor that unites our informants' opinions of the wolf zone: whether they are for or against wolves in the neighborhood, the wolf zone is seen as contentious and problematic. In the opinion of the anti-wolf informants, it places a disproportionate burden on people, while their opponents say it is ineffective at building a Norwegian wolf population.

POLICE AND THE JUDICIARY

According to many of our informants, the time and money that go into policing and adjudicating compliance with regulations are also out of proportion. People often compared investigations of incidents involving the shooting of large carnivores (usually in alleged defense of livestock or dogs) with the poor performance of the police related to other forms of crime seen as far more important. It goes to show, people said, that "animals have become more important than people," meaning basic values in society have been turned completely upside down. It was interesting to observe that this notion also invoked strong feelings in some people who are essentially positive to carnivores, in the same way it did in those who are skeptical. The topic attracted the most discussion in Trysil, where people have years of experience with police investigations of carnivore killings (mostly involving bears):

A: They pour resources into it if somebody shoots a bear. Four to five officers arrive and crawl around on all fours for a day or two—

B: Yes, Kripos [the national bureau of investigation] came too—

A: Yeah, but they haven't the time to send a patrol if someone steals all your furniture—

C: Yes, and almost rapes your kids ... It's a completely atrocious use of resources, in my humble opinion. Makes no sense. A farmer who shot a bear in Engerdal, and they send four creepy guys up from Kripos—

D: —and spend the whole summer!

C: Yes, it's become absolutely terrible! (...) [They examine the carcass to find out,] "Did he shoot it as it came toward him, or did he shoot it while it lay there?" What the hell does it matter? It looks almost like [the carnivores] are worth much more than people—than a human life.

Interviewer: Why has it become like this, then?

C: Well, a horrible outcry [from the wolf protectors], and the scientists are riding a wave at the moment. Evenstad [research center in the region], everything that can be studied and money spent on, and they've decided that this is what we really need. Illegal hunting and such, it's "*so* terrible."

Interviewer: So the scientists have power, too, since they can—

C: Yes, that's how it's turned out.

D: So what the police are told to crack down on is completely wrong, as I see it. If it's illegal, obviously they have to pull out the stops, but they need to see it in proportion to other crimes which I think are much more serious, to put it like that. It's clear they have to be resolute, goes without saying, but—

C: Well, I mean it's a completely bizarre use of precious resources and totally senseless. A lot of us feel ignored—if there's a burglary and fights and whatever, it doesn't matter, can't be bothered to send a patrol. It's almost like they're saying, "You'll have to deal with it yourself." But, on the other hand, if you're unfortunate enough to point at something carnivorous, then you're in it up to your neck.

(Hunter's group 2, Trysil)

A: There's that guy [name redacted], for example, who was attacked by a bear. Police arrived and took his boot, and it was sent off to Trondheim for [forensic] examination, to see if the rift that was in the boot matched the bite of the bear. [Much laughter.] And then [X] said he wanted his boot returned because he wanted it as a souvenir. And three or four days later, two police officers turn up to give him his boot back. You start wondering how the police spend their time. [Much laughter]

B: Yes! Look at yesterday's [TV program]: everyday crime, that's allowed now. Nobody bothers. But when there's an incident involving a carnivore, that's when the police use time and money.

Interviewer: But why would they do that? Why would the police put such huge resources into it?

C: I don't think it's the police who want to.

B: No.

C: But there are a few fanatics on the opposite side, people in favor of carnivores, including many whose work [involves carnivores].

(Farmers group 2, Trysil)

Resources spent on that kind of stuff, it's absolutely terrible, and it's obvious that people react. I can mention an example here, (...) a bear was shot up in Engerdal a few years back, to defend livestock. And I was going for an evening stroll, and two police cars appeared, with sirens at full blast, from Elverum [regional center], with I think five uniformed officers in each car—I'm laying it on slightly, but it's about right—and whizzed past at 140 to 150 kilometers an hour over here. Then I said to my wife, "Must have been a homicide and rape some place near here, must have been a dreadful accident or something." In the newspaper the next day we read that the police had been sent out to secure evidence [after a bear had been shot]. And you read in the newspaper the day after that a jeweler's shop's been burgled, and even though they have CCTV [evidence], they couldn't be bothered to send anyone. So the balance is all wrong, and it fans the flames of a conflict that's actually quite bad.

(Tourism group, Trysil)

Another aspect of the police and prosecution services' response that informants highlighted: "ordinary people" are made out to be thugs, even though they are only protecting their animals or doing things anyone do in a similar situation. It is not fair and expresses the authorities' low opinion of rural people. Several said a harsher punishment is more likely for killing a bear than for homicide, which goes to show how crazy the world has become:

I'm thinking about some of these people in the heat of a situation ... farmers and sheep owners and those, who feel that they are forced to kill a predator. They're turned into criminals and imprisoned, and for them personally it's a heavy price to pay. (...) We had one in Engerdal here recently—to think that they can't have a different system to deal with these things! There are those people investigating the scene of the crime and everything, and they can see there's been activity, lots of movement, and so they're somehow putting themselves in the situation, and then they say, no, looks like such and such happened, and the situation wasn't that desperate, so you can expect a stretch in prison, you know. I just don't get it. I'm no hunter, but I fail to understand the system—you are branded somehow. Even if your neighbors understand you, you're still known as "the one who did time for such and such." It's a heavy load for a person to bear. Because it's an animal that is killed, you get punished much more than for something else.

(Tourism group, Trysil)

We need to see this in light of the fairly widespread belief that current conservation policy is based on an urban conception of nature with little appreciation for rural people's sensible use of natural resources that has been going on for centuries. Many believe rural people are viewed with suspicion and treated with condescension. Prosecuting people who act in desperation only makes sense if seen as part of an urban conspiracy against rural folk:

If a few wolves disappear, for example, it is hardly anything but completely natural. The wolf's biggest enemy is the moose. And moose kill a lot of young wolves. That's an indisputable fact. A lot of research has been done, and they discovered a lot, so you can just talk to Petter [Wabakken][2] about it; that'll give you the lowdown. But when a wolf disappears from an area, then, that eventuality is not often raised, that it might have been a female moose that signed that wolf's death certificate, and that it might have been eaten by other animals. Then it's gone. And that creates conflicts. Because they always say it's poachers. They use that to explain why the wolf disappeared.

(Tourism group, Trysil)

This too needs to be seen in relation to the equally widespread notion that carnivore populations are larger than authorities claim and will not be much affected by the occasional shooting of an animal to defend livestock, even in somewhat vague circumstances. This perception is shared not only by opponents of large carnivores but also by some with a sympathetic opinion of them. The police's crackdown on illegal killing of carnivores did not come up in all of the interviews inasmuch as the type of data here derive from informants' own experiences and issues important to them. This means some of the pro-carnivore informants may have felt it was right to use large resources on illegal hunting even though such sentiments were not picked up in the interviews. Quite a few informants nevertheless believed the extent of illegal hunting to be significant and a serious problem.

MONITORING CARNIVORES

Electronic monitoring, today mostly by means of GPS collars, is useful for managing relatively small carnivore populations with precise population targets, according to the Norwegian Environment Agency. It can also facilitate the control of problem individuals. Managers see the technology as a way to mitigate conflict because it enables better control. But as our interviews show, many are skeptical of electronic monitoring regardless of their opinions of carnivore policy. As with other carnivore-related topics, people tend to talk about wolves the most. Some of our Trysil informants mentioned tagging of bears, lynx, and other animals, but most interviewees associate large carnivore monitoring with electronic surveillance of wolves.

An advantage of focus groups is that different opinions can be aired as part of a wider context—as integrated in a wider frame of understanding. In addition, the researcher can learn something about how and why interviewees deem various questions important by noticing which subjects appear spontaneously during the discussions and which must be coaxed out of the group (cf. the issue of "use of police resources," which was important in some interviews but not in others). When it comes to electronic surveillance of large carnivores, the informants themselves almost always raised the issue. The question was consistently associated with more general themes that many were concerned with, namely excessive or unnecessary research on carnivores, use of taxpayers' money, and the differences between wild and domesticated animals:

C: Wolves and such, it's not the same when a wolf comes along with a huge [radio transmitter]. It's kind of not the same. Is he wild or is he a domestic animal? You don't get the right feeling for carnivores—with those hideous transmitters and whatever else they're walking around with.

A: All organizations and scientists, anyone in fact, who's fond of nature—we hunters are probably much fonder of wildlife and nature than some of those guys, I'd say. Nevertheless, there's no stopping them with their nasty transmitters and implants in the stomach (...) and stuff, you know. It's positively disgraceful! There was this time I caught a fish in the river, and it was tagged. They'd shot [the implant] as far down inside it as they could—an enormous festering sore! I don't know whether it would've survived. It looked really nasty. Imagine how a larger animal must feel, with that slab around its neck. It's playing around with nature, I think. Interfering so that someone can have fun at work and earn money, and all—total waste! (...) It's my money as well. I pay my taxes. (Hunters group 1, Trysil)

For anti-wolf informants, opposition to electronic surveillance is part of a general pattern of criticism of what they perceive as interference by politicians and scientists in traditional land use rights. Hunters, farmers, and landowners see themselves as the rightful stewards of the land, since the physical environment is integrated in the practices that carry their life projects. In their eyes, the government has demonstrated beyond any doubt that it lacks the ability to sustain rural livelihoods. First, the cultural landscape is deteriorating, partly as a result of reduced opportunities to use grazing land in areas with carnivores.

Second, it has deprived local communities of the right and ability to protect livestock, dogs, and valuable game from predation by protecting carnivores, especially wolves. For these informants, electronic monitoring is a concrete reminder of who controls the natural environment, and it is not them. They are excluded when information is collected and collated and when decisions are made. It is therefore logical for them to ask what all the research is good for:

> This research on bears, for example—what is it they don't already know and have to find out? If they've still not managed to figure out how far a bear travels after doing research on them for twenty years and radio tagging a large number, it's beyond me. Their work has obviously never been evaluated. It's as if I kept on building a house forever.... And it's not the same feeling you get of the natural environment either, if you're lucky enough to see a bear—with a huge wedge around its neck.
>
> (Hunters group 2, Trysil)

These excerpts are from focus group sessions with hunters, most of whom experience nature and wildlife as the centerpiece of their hunting practices. They believe electronic carnivore surveillance corrupts this experience. Those more concerned about farming and landowner interests maintained that large carnivores do not belong in Norway today. Also on this score, the wolf attracted the most attention. The Norwegian wilderness has gone, and the wolf—the incarnation of the wild—will have no place in the used cultural landscape they believe is characteristic of Norwegian nature today. That the surveillance of wild animals is necessary was seen as more proof that Norwegian nature is no longer wild and thus that large carnivores do not fit in. In that sense, and in light of indications of a corrupt or partly domesticated wolf population, some felt you may as well take the situation to its logical conclusion and fence the wolves in:

> When it comes to fencing and stuff like that, it would have been much better, given the huge spaces we have, to fence them in. And then we could manage the carnivores in relation to inbreeding problems, and control most of [what happens] between the different species—inside the fence. There would be no conflicts with landowners and livestock farmers. [The situation] would be much more organized. It would be a much better political strategy in my view, both management-wise, in relation to conflicts, and at the local level.
>
> (Neighbors group, Trysil)

As far as wolves are concerned, it is important to remember here that the anti-carnivore interviewees regularly differentiated between "real wolves" and our current population (as described in chapters 5 and 7). Like the more carnivore-friendly informants, they often talked about the animal itself using positive terms, while stressing that what they said—unlike the friends of the carnivore—applied to the species, not local individuals they saw as utterly out of place (see also Figari and Skogen 2011). For those anti-carnivore informants who expressed admiration or respect for the wolf itself, their ideas about the wolf are closely tied to what they see as the animal's natural habitat: the wilderness. This clashes with their image of the local environment as used land, where even the forest is seen as a cultural landscape. Accordingly, the local wolf is considered an impure animal with all the signs of human corruption, since it lives in a landscape that is not "wild." People saw the need for monitoring as evidence of Norwegian nature's unsuitability as a habitat for wolves today: in the wilderness, where wolves belong, they could and should roam free without interference from humans.

Many people with sympathy for large carnivores were also critical of electronic tagging. Like many of the hunters, they saw tagging as incompatible with the wild nature they would like to see in Norway. Unlike their opponents who saw local landscapes as used land, the pro-carnivore informants preferred to see the same landscape as wilderness where carnivores have a natural place. By their mere presence, carnivores convey a sense of the pristine and unspoiled. The sight—or even the idea—of large carnivores with collars detracts from this experience. For these informants, too, electronic monitoring contravenes the image of what a wild animal is and should be. One of the informants in a neighborhood group, who lived within the home range of a wolf pack and appreciated having wolves in the vicinity, said: "I'm against tagging. Tagging, it's not a good thing, in my view. (...) Wolves should roam free. Otherwise they should fence them in. Then they would know where the wolves are [ironic]."

Some of our informants raised the ethical side of electronic surveillance, a concern shared as much by anti-carnivore informants (see the first quote in this chapter) as pro-carnivore ones (see also chapter 5, where several interview excerpts illustrate the issue). All the same, the welfare of the individual wolf was not central to either group's criticism. As shown in chapter 5, antagonistic attitudes toward wolves in Norway are not the result of divergent interpretations of the wolf's nature. On the contrary, pro- and anti-wolf informants appear to agree on what kind of animal the wolf is. Of all the characteristics people attributed to the wolf, its "wild nature" stood out as particularly significant. It is in this light that criticism of electronic surveillance should be understood. Physical signs of human interference, such as electronic collars, challenge the idea of the wolf as the wild incarnate, of the uncorrupted. Regardless of whether one wants wolves in Norwegian forests, this is the crucial idea—self-evident, intuitive, and thus more commanding. At stake is a very important distinction, namely, between what is conceived as wild and what should be subject to human control and manipulation (see also Figari and Skogen 2011).

Nevertheless, some informants said they understood why it was necessary to monitor large carnivores. They cited official carnivore policy that necessitated in their view an extreme form of regulation, particularly for wolves, since parliament decided exact population goals. It is hardly possible to fulfill such detailed requirements without monitoring individuals. However, only a minority of our interviewees referred explicitly to the policy framework, which, as we saw in the section on zoning, many did not consider legitimate in the first place. Despite the widespread criticism of tagging, many felt the wolf population must be kept under control. On this point, too, the views of anti- and pro-wolf informants tended to converge. Conservationists wanted a larger wolf population in Norway but conceded that this (much) larger population would have to be controlled in the future. Many anti-wolf informants believed that if we are to have wolves at all, then the government should keep them under control—and that this is needed now.

So on the question of electronic surveillance, as on other carnivore-related issues, opponents and champions coalesce into a kind of united front in their critique, although they normally do not recognize this themselves. That many who speak critically of monitoring also press the need to keep carnivore populations under control may seem paradoxical. One might call the fact it a weakness of our informants' argument against electronic surveillance that population control necessitates some form of monitoring. Many informants, however, had great faith in local knowledge and corresponding distrust of knowledge produced and disseminated by scientists and government officials (as described in chapter 6), not least on population estimates. Opposition to electronic surveillance must therefore be understood as part of our informants' general suspicion of plans and interventions that detract from what many perceive as their right to utilize natural resources that belong to them. More generally, it revolves around questions of knowledge and power we have already discussed in more depth.

COMPENSATION

Compensation for loss of livestock, and in certain circumstances of dogs, is a pillar of the Norwegian approach to carnivore management—as in many other countries. Awareness of compensation schemes is naturally greatest among the people most likely to need them, that is, livestock farmers. The amount of compensation paid out per lost animal seems reasonable enough in the opinion of most informants from the livestock sector:

> Interviewer: What about compensation schemes then, how they work today, is it—?
>
> A: If you can [have it verified], it generally works pretty smoothly. They deduct what they think you'd have lost through natural causes and—

B: I think we're in a special situation in Trysil. We've had depredation problems for over twenty years, and [the authorities] are aware of the problems around here. It's much worse in other places. When there is not much that can be verified and the problem is relatively new, it can be a much bigger problem than round here.

(Farmers group 2, Trysil)

Criticism of the compensation system is more indirect. The payouts do not cover additional work involved in heightening supervision or the time and energy spent looking for carcasses. Loss of breeding animals is not compensated fully because the rate is the same for all lost animals, and payouts can never replace high-quality breeding stock. A lot of the effort that goes into improving the breeding stock is lost forever when important breeding animals are killed. Compensation cannot safeguard the use of grazing areas in the longer term because the emotional stress and additional work needed to look after the remaining animals are so demanding that to continue is impossible, and younger people will shy away from this form of production and may not take up farming at all.

We asked sheep farmers what they thought about the Swedish compensation model, which pays people for having carnivores in the vicinity, regardless of the losses incurred. The model is designed to promote prevention, since it pays to simultaneously deliver animals to slaughter and collect compensation, as this model provides for. In Norway, a similar program was proposed in a 2003 parliamentary white paper on large carnivore policy, partly as a supplement to and partly in place of the present system. However, the center-left coalition government that came into office in 2005, which included a party with strong ties to agriculture, immediately scrapped the idea:

A: We don't think much of that model, at least personally I don't.

B: Well, it's a typical armchair idea—

Interviewer: [It is implemented in Sweden, but for reindeer husbandry and Sami communities.]

A: Yes, I reckon it's OK for the reindeer industry [which is organized in a completely different way than sheep farming in Norway]. Because in Sweden you won't find hardly any sheep grazing on uncultivated land, we're talking about microscopic numbers.

(Farmers group 2, Trysil)

Our farmer informants rejected the idea for several reasons. People with the highest losses would get no more than those with the lowest, so the scheme is fundamentally unfair. And agreeing to this type of scheme would amount to capitulation because the government would be paying them to agree to permanent populations of carnivores and to letting carnivores dictate how they should run their farms—perhaps forever. The sheep farmers could not accept carnivore populations so large that this would be necessary and thus rejected this type of scheme. Compensation for actual loss is compensation for lost income and therefore right and proper, and it does not require farmers to change how they run their businesses. It is a lesser evil, as farmers' ways of farming should be allowed to continue unchanged—if anything, the carnivore management regime needs changing. They claim historic grazing rights entitle them to rough grazing without incurring major losses due to carnivores, and they therefore have the right to demand that carnivore populations are reduced so that these rights are protected. Other forms of compensation are still discussed, though this should probably be seen more as an expression of desperation—where the carnivore situation is seen as part of the accumulated threats regional agriculture is up against—than as a practical political proposal:

A: Then (...) the government should rather say: we'll get rid of the sheep in Trysil and only keep the large carnivores. Then they can buy us out, and we quit.

B: Yes, that would the most straightforward solution, in fact, even though we would probably have been just as angry and aggravated if they did, but at least we would have known where we stood, instead of this torture.

C: Yes, I'm with you on that, but compensation shouldn't follow the person, it should follow the property.

A: It could be worth a great deal for many years to come.

B: Can't agree more.

C: Because if you buy me out, what about those who will take over the farm? It has to follow the property.

D: It would be much more honest, too. If they said that east of [the river] Glomma, that's where bears and wolves can live, but no worry, we're going to buy you out, and we'll be giving a decent amount to anybody who's affected. It's really like expropriating your land. It's the same thing that's been happening already in practice. I mean, you remove somebody's rights [because we can't graze our sheep freely even now]. But then you should be paying an annual compensation, because it wouldn't work otherwise.

(...)

B. We can see it from a slightly different angle, we've been enormously privileged to take over a farm, and we've got a duty to pass it on. It is a legacy we are given in trust and which in some way or another we're going to hand over to the next generation. We've been given a responsibility as stewards [of the land].

A: Yes, because when we hear about all the problems we have to contend with, it's a peculiar thing—

B: —that anyone [will go on]!

A: —that anyone has the energy to do it, plain and simple.

(Farmers group 2, Trysil)

Not surprisingly, some of our informants sympathetic to large carnivores saw both the compensation schemes and farmers' morals in a different light. Farmers and reindeer herders all try to squeeze whatever they can from overly generous compensation schemes with very lax requirements for documentation of depredation. One of the Østfold neighborhood groups used a recent media report as an illustration:

A: Then there's that business with compensations. It's a cock-eyed system if ever there was one, you don't even need to produce proof. There is so much fraud. It explains why there's so much hostility between people in favor of large carnivores and farmers and hunters and suchlike, because we don't trust the farmers, because they fabricate the figures they report [to the authorities]. The wolf gets the blame for everything, everything—almost everything at least.

(...)

B: They're obviously doing well, in any case.

A: We saw in today's *Aftenposten* [daily Oslo newspaper], forty-four thousand reindeer had been killed by carnivores just in [the county of] Finnmark [according to reindeer herders].

C: Wasn't it thirty thousand that had been taken by golden eagles? They said so on the radio today.

B: Well, it certainly meant that every golden eagle must devour two reindeer every day. [Much laughter]

A: So what it adds up to in the end is that we can't trust the government either, when they accept this!

(Neighbors group 2, Halden)

Some hunters were aware of the government's compensation scheme for hunting dogs, but many hunters who owned dogs knew little about the compensation criteria. Those who did know the scheme felt it was too limited, that is, it covered too few situations (only hunting and training, not visits to the forests for other purposes, farmyards, etc.):

A: Doesn't make sense. There's nothing called self-defense [defending the dog] or common sense anymore. In [our area] we have a female wolf that has taken nearly fifty hunting dogs, and they can't get permission to shoot her. She's even gone up the steps and taken dogs tied to the porch, on the farms. Doesn't seem to matter. She even made a grab at a dog the owner was out walking on a leash.

Interviewer: Starting to get pretty aggressive, then.

A: Well, you never know where the line goes for an animal that's put its mind to it, to put it bluntly.

(Hunters group, Halden)

Some of the people who had received compensation for lost dogs had positive things to say about the scheme and the officials representing it:

Interviewer: So how did the official act? Was he a likeable guy?

A: Yes, very nice! Very nice! The local paper sent a reporter, among other things, and I filed [a compensation claim]. It feels good to get compensation for something like that, and I thought that this time the government could pay because they're always trying to get as much as possible out of us farmers. Everything we deliver is falling in price, you know ... I posted my claim on the Monday and had [the money] in my account by Thursday. You won't get service like that from the County Governor for anything else!

(Farmers group, Halden)

In contrast to this dog owner, others found the compensation rate too low for really good and top-of-the-range, prize-winning gun dogs; in no way does it compensate the dog's real value. People often spoke of the scheme in connection with the right to defend the dog, which at the time of our interviews was not in place (that has changed since, so people can now shoot carnivores actively attacking dogs during hunting or training). The inadequate compensation scheme and lack of a right of self-defense were seen as evidence of the government's nonchalant attitude toward hunters, compared to the attention sheepherders and conservationists receive:

A: Since we're talking about dogs and wolves and stuff, apropos the law, the fact that there's no right of self-defense when it comes to dogs, it's totally perverse in my opinion. It's OK to shoot a wolf for attacking a lamb. I don't know how much a lamb is worth, but if a wolf takes a champion gun dog worth about sixty to seventy thousand kroner [USD 8,500–10,000], you're not allowed to shoot at the wolf when it attacks your gun dog. There's a great many hunters—clearly most hunters would probably take the law into their own hands and shoot the wolf, at least I would.

...

B: The problem is they're so crazy about protecting these carnivores. There's never room for doubt. (...) It's always for the benefit of the carnivores!

(Hunters group 2, Trysil)

Loss of dogs was not particularly relevant in Aurskog-Høland (one of our study sites; see chapter 2) in 2009. There were no wolf packs in the municipality at the time, and according to our informants, several years had passed since the last time a wolf attacked a dog. Nevertheless, the issue provoked strong emotions. Mistrust of authorities was an undercurrent; there was no point in expecting to get anything from them. Furthermore, loss of dogs could not be accepted under any circumstances, and the compensation schemes are therefore not what people are mainly interested in. Shooting in self-defense is another matter:

A: No skin off my nose to say that if I'd seen a wolf devouring my dog in the forest, then I'd—I don't know what I would have done.

B: I've had the same [thoughts], and I don't know what the outcome would be. It's something you need to decide there and then I think.

A: The Swedes have introduced self-defense for dogs, haven't they?

B: Mm.

A: Well, it shows it's absolutely necessary here, too.

(Hunters group, Aurskog-Høland)

A study from Wisconsin (Naughton-Treves et al. 2003) showed that compensation for lost hounds (used for black bear hunting) had had no effect on owners' attitudes about wolves. At the time of the study, the Wisconsin scheme did not compensate all hunters whose dogs had been taken by wolves. The scientists compared attitudes toward wolves among hunters who had received compensation and hunters who had not and found no difference. The vast majority felt there were too many wolves in Wisconsin. This seems to suggest that compensation schemes, considered an obvious right in light of the government's policy on large carnivores, do not affect attitudes shaped by other, more fundamental factors.

"CARNIVORE-PROOF" FENCES

The government supports a range of efforts aimed at preventing livestock depredation by large carnivores. This includes early retrieval from summer grazing, keeping livestock in infield pastures, and installing "carnivore-proof" electric fences. In the parts of the country where we did our study, all these measures were implemented. Early retrieval and infield grazing are resource-hungry options but only affect the farming sector itself. We therefore decided to focus on a measure that also affects other land users and possibly wildlife species other than carnivores: electric fences designed to deter carnivores. One such fence was being erected in Trysil while we were doing our interviews there. Unfortunately, however, we have no interview data on this issue from the time after the fence was installed. This is particularly unfortunate because the first grazing season proved hugely problematic. Several lynx operated inside the fence, and a bear found its way past it as well. As this was a radio tagged female bear, which could possibly provide several reproductions on Norwegian soil, a helicopter was brought in to relocate it. No permit to cull the bear was issued, even though it was inside the enclosed area together with the sheep. This decision upset a lot of people. At the time of our interviews in Trysil, construction of the fence had just started. A road alongside the fence was also being built. The whole plan was discussed in the local media and loomed large on the local political agenda. People were therefore familiar with it, and we were able to obtain opinions on the strategy in general and the likely consequences for the area where the fence was being built.

To judge from reports in the local newspapers (primarily *Østlendingen,* based in the regional center Elverum), letters to the editor, and online debates, carnivore fences continued to evoke strong reactions after we left Trysil. Understandably, problems related to the fences' functionality received the most attention. And this shifted media attention from the Trysil fence to a similar one built in the municipality of Grue, a little further south. In the summer of 2010, several predators of different species massacred a flock of sheep inside the enclosed area. Rumors of sabotage by wolf lovers were countered by allegations that the fence was purposely faulty to prove that carnivore fences were ineffective. However, both fences have had a good record for the past few grazing seasons, measured in lost sheep. Although we have no data on how people view the fence in Trysil today, what we learned in 2007 and 2008 shows opinions are rooted in more fundamental views on large carnivores and wildlife management generally.

Some facts about the fence in Trysil: in the summer of 2008, six sheep farmers used an area of 21.5 square kilometers for the first time to graze 914 sheep. The fence is 120 centimeters high and has five high-voltage wires. A road runs alongside the 23-kilometer-long fence and is used for checks and repairs. Most of this road was built while the fence was being put up. Outside the grazing season, the power is turned off and the fence is laid horizontally onto the ground so as not to impede movement of people or animals. While the fence is up, it constitutes a barrier to both humans and larger animals such as moose. The Hedmark County Governor has paid 1.7 million kroner [USD 250,000] for the project and for the time being will continue to support operations and repairs with an annual sum. While the fence was being planned, opinions in Trysil were divided. The large and powerful Trysil chapter of the Norwegian Association of Hunters and Anglers was highly skeptical of the idea because of the consequences for wildlife, hunting, and general access. The farming sector was relatively positive to the idea, although they considered it an emergency antidote they would rather have done without.

When we did our interviews, planning and early work on the fence had started and the project was well known, at least in circles that used forests actively. Not surprisingly, we found a wide variation in opinions, though a distinct pattern did emerge. Support for the fence was (understandably) strongest among livestock farmers themselves, while hunters and other important groups of land users were unhappy about the fence and expressed some very negative opinions. Sheep farmers were particularly frustrated given the carnivore situation in Trysil and loss of livestock. Several saw the fence as a last resort in a continuous struggle for the right to graze sheep. One sheep farmer said, "I'm doing everything I can to get my sheep into the secure area in Trysil; if I don't then it's [over] for me." At the same time, sheep farmers were determined to stay in the business and were not at all thinking of giving up:

> I'm one of those that hope and believe [things will get better with] the electric fence. It'll be like the dawn of a new age, as I see it. Hopefully the carnivores won't get in. But my strongest feeling is one of relief because the [sheep] won't be wandering into populated areas and causing problems. (...) I don't think livestock farming has any future in areas where [the government has decided to have] large carnivores. Some politicians believe it has. I've seen some brochures from Jehovah's Witnesses picturing a child sitting next to a lion which he's cuddling, but [laughs] I don't think the world's ever going to be like that.
>
> (Farmers group 2, Trysil)

One sheep farmer we interviewed was unhappy with the fence project and had no intension of using it. He mentioned certain features of the project that in his opinion were problematic:

> What the others are relying on, this carnivore fence, I'm not happy with at all. It might be OK as long as the project is running, but what about afterwards? Who's going maintain it? Are we going to be responsible for taking it down and putting it up again? And not only that—now that we've had our animals in the vicinity—we're constantly out in the forest [looking after them], bringing sick animals home and treating them and getting them out again. That'll probably be impossible because [the fenced area is so far] from where we live. So I don't think it's a good idea. It would actually be better to quit. And that's a very bad solution, for the community where we have our land will be the poorer, because everything will be overgrown. You can say we're not dependent on these sheep. We can live without them, but the environment around us will be the poorer for it.
>
> (Farmers group 2, Trysil)

Other informants supported the fence, including people in tourism. There are large areas in Trysil for outdoor recreation, they said, so the fence is unlikely to cause serious problems for tourism. But they acknowledged that it represents a serious interference with nature and a symbolic—if not physically absolute—barrier to people who use the land:

While the [enclosed] area is relatively large, it's not *that* big. But it will be a strange situation, with mostly a single species grazing there. I reckon the moose will vault over the fence easy as pie, but a few other [species] won't get through, unless they are small enough to crawl between the wires. People will be shut out too, in a way, because you might not want to cross a fence. The land inside is privatized in a way.[3]

(Tourism group, Trysil)

Participants in most focus groups appreciated the situation sheep farmers found themselves in and their need to protect livestock from large carnivores. Nevertheless, most were worried about the negative effect of fencing in a large forest area. They criticized access restriction (physical and symbolic, not legal) and the impact on ungulates and other wildlife. Not many believed the fence offered sufficient protection against carnivore attacks. And it was unreasonably expensive, not least in view of the subsidies transferred to the sheep farmers already. The strongest criticism concerned the aesthetic and ethical aspects of interfering with nature. Before the 23-kilometer-long electric fence was built, trees were cleared and a road built around the whole area. Many informants were upset about the "destruction" of the land:

I'm amazed the authorities allowed such colossal interference in the natural environment. I know, as a forester, how difficult it is today to get permission to build a logging road or a tractor road, and here the ground obviously has to be leveled where the fence is going up. That they're allowed to interfere with nature to that extent, what with the restrictions on the logging indus-try, I can only say I'm surprised. It seems, to put it bluntly, like some sort of overcompensation.

(Neighbors group, Trysil)

This is from an interview with outspoken carnivore opponents and an ex-ample of the lack of correlation between negative attitudes to carnivores and positive perceptions of carnivore fences. Many hunters and farmers have similar views on the presence of carnivores, but their reasons for concluding that large carnivore must come down often differ. Although "hunters" and "farmers" are di-verse categories, which also to some extent overlap, we see a pattern in which two different types of practices connect to different conceptions of the ideal land-scape. While many farmers we interviewed wanted to see forestry, farms, and livestock farming around Trysil, where the legacy of generations of human activ-ity adds to the value of the landscape, hunters often talked about a landscape in which their sense of contact with nature and wild animals is the central feature.

These different conceptions of the landscape are described here as "ideal types" insofar as we present them in a simplified but clear-cut manner (Sku-land and Skogen 2014). In reality, no distinct boundary exists, and many will find these interpretations of the landscape familiar. The point is that a conception of how the natural environment in Trysil is today and may become in the future has a crucial bearing not only on how people view the presence of carnivores in the area (see chapter 5) but also on their attitudes to electric fences in the forest. Both livestock farming and hunting provide foundations for landscape interpretations with limited room for large carnivores, albeit for somewhat different reasons. For sheep farmers, carnivores present a threat to livestock grazing and the cultural landscape. For hunters, carnivores threaten game stocks, and, when it comes to wolves, is a problem for those who hunt with dogs.

So while sheep farmers and hunters often share an opinion of carnivores, we see from our interviews that carnivore fences are not compatible with hunters' interests or their conception of the landscape. In line with what has been said about the importance of experiencing nature and wild animals, hunters unsurprisingly proved particularly critical of carnivore fences. The aesthetic aspects of the fence project upset them, and some hunters were concerned about the impact on the moose stock and other wildlife populations. The excerpt below reveals a general skepticism about how suitable sheep are for rough grazing and tells us that the disagreement between some hunters and sheep farmers runs deeper than the controversial fence issue (as discussed in chapter 5):

> A: Migration routes of many animals will be changed completely.
>
> B: A small moose calf won't get over this fence.
>
> C: Well, the fence is going up in June, middle of June, if it goes according to plan. OK, so a few calves have been born inside [the closed-off area]. What's going to happen to them? The mother can jump over. But then you've got the calf, poor thing. Many calves get trapped in the ordinary sheep fences. Will it get trapped, then, in the electric fence? It's pretty serious, I think.
>
> D: It's not [very likely] it will get stuck—
>
> C: No, but it could happen.
>
> D: Yes, it does happen.
>
> C: Or even a sheep. Sheep, they get stuck in anything.
>
> A: You won't find a more stupid creature [than a sheep].
>
> (Hunters group 2, Trysil)

Most of our focus groups expressed a great deal of skepticism and gave many reasons for distrusting the fence project. For example, an official working for Trysil municipal council had no faith in the effectiveness of the fence:

> I know for a fact that bears can jump over ordinary fences of normal height, carrying a sheep, as we saw last summer. [A farmer] had fenced in his sheep on the summer pastures, and [found] a carcass. And the bear had found a way

in and taken the sheep clear over the fence without leaving a tuft of wool or anything on the fence. It had just lifted the sheep and taken it over the fence and up into the forest. So it's not as easy as just erecting a fence either, I'd say.

(Trysil municipal council officer)

With the exception of some of the farmers we spoke to, people both for and against having carnivores in the area felt the fence and the new road would have a serious impact on the natural environment. Both sides want to see nature preserved where they live, although they may have different ideas of what the natural environment should contain and who should manage it (see also Figari and Skogen 2011; Skuland and Skogen 2014). Carnivore opponents, once again, have found common cause with people of a more sympathetic inclination toward carnivores:

That fence irritates me no end, just so I've said it. It was the only pristine wilderness area we had left here, and now they've fucking ruined that too. They've dug roads across the entire beautiful hill between here and [the next valley]. It's totally destroyed. Where were the environmentalists then? To think that something like that is allowed! I just don't understand—that lynx take sheep, that's too bad of course, but I think this is absolutely horrendous. They've razed the entire area. Then they're going to lay the wires down, five wires, laid down [poles and wires], on the ground, and it'll trap game and dogs, you know. Now it's actually illegal to have wire fences lying on uncultivated land that's not in use, you know, you have to remove them. And then they're pushing ahead today with something like this. Completely meaningless in my opinion. But, in ten years, just you wait, it'll be gone. Because it won't work. At least, I don't think it will. But by then the land will have been destroyed, the roads they're building they're enormous.

(Hunters group 1, Trysil)

HUNTING LARGE CARNIVORES

The hunting of carnivores is often assumed to mitigate some aspects of the conflict. For example, a parliamentary white paper on large carnivores in Norway[4] stated:

There are clear indications that culling and hunting of large carnivores have a conflict-mitigating effect. This appears to be the case in Norway and in several other countries with strong hunting traditions, perhaps particularly in Eastern Europe and Sweden. This is partly because hunting and culling limit population growth. Culling and hunting with local participation can increase the legitimacy of the management regime. It is also a common experience that hunting and culling make large carnivores more shy of humans. (p. 26)

We agree, based on what several hunters have told us and on our own observations of the trends in conflicts over species hunted today (in both Norway and Sweden), there are indications that hunting large carnivores may have a beneficial effect on conflict levels. In addition to the reasons presented in the white paper, hunters we interviewed also highlighted the excitement and attraction of the hunt along with the need to perform predator control, following from their perceived responsibility as stewards of the land. Nevertheless, we can discern a hierarchy among the arguments put forward. As many hunter informants were at pains to stress, the main reason for hunting large carnivores is to control populations that would otherwise get out of hand. But when the "stewardship" box is ticked off, that hunting itself is exciting and challenging is certainly no disadvantage; it nourishes a sense of mastery, command, and perhaps the respect of other hunters. Some forms of hunting are new to Norway, so according to our informants, to develop hunting methods is both stimulating and satisfying. Here from a discussion about lynx hunting:

> A: It's an unbelievably fascinating animal to hunt. She passes through ... here at night, and if we find a track, then it might [move long distances]. Now, we're right up against the Swedish border round here, and it often wanders over there, so then we just call it a day. Otherwise, if it moves [into other hunting districts on the Norwegian side] we'd be telling people, "Look out, the cat is on its way!"
>
> Interviewer: Fascinating because it's so intelligent and—?
>
> A: Yes, sure.
>
> B: It's special, you know. If you track it while hunting ... it can come and go right behind you, that's a fact. It's extremely wary, but it's not that shy either. It's in control, all the time.
>
> (Hunters group, Aurskog-Høland)

> A: We started last winter. Or maybe the year before, but one was shot in eastern Østfold, or was it [just across the county line] perhaps?
>
> B: It was a [female], so they stopped everything [i.e., the hunt because the quota for females was filled]. But last winter we made a new attempt. It was extraordinarily popular. You can round up sixty men in half an hour flat, no problem, middle of the day, weekday, whatever. (...) A message was sent out; we were out eight times. At the lowest we were never fewer than thirty, and at most there were seventy of us. Obviously, it's new and popular, like things are in the beginning, so we may not get the same response this year, they might feel it's getting boring, because there's a lot of waiting, and it's cold, and—
>
> A: But don't you think they understand the importance of it and—the need?
>
> B: Reckon they do. A lot of people do.
>
> Interviewer: So you think that's the main reason then? They're aware of the need to—

B: Yes, most of them are, at least the [most active] among them; they see it as a community service. And there's a large number of hunters. But, as I say, those hunters who enjoy hunting [lynx], those are the ones we rely on.

(Farmers group, Halden)

Some hunters suggested that hunting wolves might make hunters less averse to wolves, though it was not a predominant viewpoint. These passages illustrate the range of views:

Interviewer: What about the wolf? They'll be introducing hunting permits sooner or later. Is it—?

A: I'm so stupid [ironic] that I think that it will be popular with the hunters.

(Hunters group 1, Trysil)

A: And as far as wolves are concerned, I think stray animals will be accepted, but many object to the permanent family groups here. But I think maybe people's notions of these animals might change if [the government allowed] licensed hunting, or some other form of hunting, and wolves became game. Having [wolf] populations would obviously be attractive. So it should be allowed [as soon as] it's feasible.

(Neighbor's group, Trysil)

Interviewer: But licensed hunting is on the cards, at least in Hedmark; they had it before, licensed hunting of wolves, and it'll almost certainly return— that's an exciting prospect, isn't it?

A: Well, it may be—

B: Yeah, would be very exiting, for sure!

Interviewer: But not as exciting that it somehow outweighed the disadvantages of having wolves in the area?

A: No, it wouldn't, in my view at least.

C: If licensed hunting were to be introduced, the size of the wolf population you would need for that to happen would be a problem in itself, and it would harm the populations of moose and roe deer ... No.

Interviewer: You [who live] inside the wolf zone may have a long wait before licensed hunting is introduced, that's clear enough.

C: Pretty far into the future, true.

(Hunters group, Aurskog-Høland)

So while hunters appear interested in hunting wolves, it is not without ambivalence. Statements concerning possible changes in attitudes to the wolf should be construed not necessarily as applying to the informant's own attitudes, but rather as generalizations of the same type expressed in the white paper on large carnivores. That the possibility of a change in attitudes is acknowledged in some hunting circles is nonetheless significant.

As for hunting large carnivores, lynx hunting has the longest history in our study areas and the highest number of participants by far. In reality, it is the only form of hunting of large carnivores of any appreciable scale in Norway. Hunting bears and wolves is recent and very limited, while wolverine hunting—more prevalent and important in other parts of the country to reduce livestock and reindeer depredation—only takes place on the margins of the area where we conducted our studies. Lynx hunting is therefore the main form of large carnivore hunting, and hunters' experience could be important for assessments of the possible effects of licensed wolf hunting. We conducted two focus group sessions with lynx hunters we might call the core members of two different hunting teams from two different parts of the study area. Other groups had informants with experience of lynx hunting as well. Taken together, we have quite a reasonable amount of data on lynx hunting in southeastern Norway. A couple of informants had been involved in licensed wolf hunts, and some had been bear hunting in Sweden (done on a much larger scale there, given its much larger bear population), a few on a regular basis.

What concern hunters the most are without doubt quota sizes and how they are set. The hunters commonly believe quotas are too small and widely assume the lynx population is substantially larger than official estimates say. The monitoring method is faulty, informants say, and wildlife managers are unwilling to accept observations not made according to official procedures (as pointed out in chapter 6). Their own observations are the most reliable, our informants claim, and the number of lynx they believe live in their area is regularly at odds with wildlife managers' much lower estimates. These population figures, along with the conservation ideology presumed to pervade wildlife management, explain why quotas are far too low.

Moreover, some were discontent with the division of hunting districts and their allocation of quotas—again, mainly because quotas are so small. Hunting teams in some regions have to compete for only one or two animals. The quota is filled (or overfilled) in a very short time, and some hunting teams get no lynx at all before the season is closed, which happens automatically when quotas are filled. This is often interpreted as evidence of abundant lynx populations; otherwise, hunting teams would not be able to find the lynx as quickly as they do:

> A: As it turns out, well, it's pretty much hit and miss; hunting conditions have varied widely. We're talking about lynx here, so we need snow, naturally. And like it is now, when we had ideal [snow] conditions all over the region, from Nes [here in the north] and to Halden, and it was only this year they decided to allocate one of the two lynx we could shoot to Halden and Are-

mark [next-door municipality], (...) we've mostly shot lynx up here, because we have had snow, and it's rained down there. So they've been complaining a bit, I guess.

C: Yes, that's [what we heard]

A: But as it turned out, from what I heard on the radio, when it really did start snowing, [then I assumed] these lynx were shot—I just stood like this [with the radio] in the evening, because it was evening [when it started to snow], and just waited. We knew then the lynx hunt would be over the next day. That's how it is. There are so many lynx. (...) No, the lynx quotas around here are not too high, that's a fact.

Interviewer: Just two animals on the quota—how many municipalities are we talking about?

B: Yes, how many is it? There's Halden, Aremark, Id, Marker, Rømskog, Aurskog-Høland, Nes, Blaker—

C: Uhh, Blaker, are they with us? Are they? Nah.

A: I don't know, it's on the Internet. Should have liked to have seen the dimensions, how big the [area] was. Aurskog-Høland, how big is it? [Six hundred square kilometers?]

B: Even bigger, I think.

(Hunters group, Aurskog-Høland)

Seen from a management viewpoint, procedures for reporting shot lynx must be as accurate as possible to prevent overshooting. Since quotas are small, the hunt must stop the minute the quota is met. Hunting teams therefore must follow how the hunt is progressing so they can quit at short notice. But, our informants argued, mobile reception is often poor out in the forest, and requiring people to call in regularly is too rigorous. Underlying this notion is probably first and foremost dissatisfaction with quota sizes, not that it is particularly difficult to stay updated. The small quotas necessitate strict rules on reporting and maintaining contact. Some also pointed out that strict enforcement of the rules necessarily involves the criminalization of ordinary hunters, who have no desire to break laws or regulations. But "to hunt against the clock," is difficult and perceived as contrary to the nature of hunting and ruins the pleasure:

B: It is too bad that there is this great pressure to get it all done quickly.

A: Like having to shoot them at a certain time of day, like, it's just stress and strain. It puts a damper on it, compared with what it was like before. And that's what makes it necessary to organize it very well, because you only have the one day. It would be great to have success on the only day of the hunt.

D: Yes, it's not even a day any longer.

B: It lasts to two o'clock.

C: Yes, it lasts until two nowadays, and then you have to [call the County Governor's office between one and two]. And that's when it starts getting slightly crazy—it's wrong.

B: You can't hunt against the clock, somehow; it's not possible. No hunters are used to doing it. (...) You've been at it for days [tracking and preparing before you are allowed to shoot], and then you've sort of got an animal [under control], and then if you shoot it at five to two or if you shoot it at three o'clock, I don't see it [as a problem].

C: Why do they want to criminalize people, I mean, turn ordinary hunters into bandits? After all, ordinary people don't want to be crooks, but they're doing whatever they can to make a crook out of you. It's completely tragic, in my view.

(...)

Interviewer: Do you think [the price for taking part in the hunt] is reasonable, or—?

A: I mean, it's only a few hours.

D: Fifty kroner (USD 8) an hour!

C: It's the most expensive [form of] hunting! [Laughter]

A: It's not that—

B: Nah, hundred kroner won't hurt anybody today.

A: But it's the principle, hunting is so darned regulated, we won't be able to hunt at all soon. Can't even hunt for a whole day.

Interviewer: And if you don't get the lynx, you still have to pay up.

(...)

A: We would still pay the three thousand kroner. And they could have called us, maybe nine in morning, before we'd started out: "A lynx was shot in Åmot, so it's all over, guys. You've paid your three thousand kroner [for nothing] and you're not getting a day's hunting even."

(Hunters group 2, Trysil)

Small quotas often create competition between hunting teams in the same region. If one or two teams get a lynx, the others go home empty-handed. It's about shooting the lynx as quickly as possible, so you need to know where it is, which is why a great deal of intensive tracking precedes the hunt and why hunting teams need to be so large. You need a lot of hunters to get a lynx within an hour or two, which is often as much time as you will have before your neighbors bag their animal. A side effect is that hunting comes across to the outside world as a form of methodical killing, where the lynx's chances of survival are nil. This perception may give lynx hunting a bad reputation, and some fear it may hurt the general public image of hunting. That risk, according to some informants, is especially high because the particular characteristics of lynx hunting also attract widespread media coverage, which frequently shines the spotlight on the negative aspects of such a massive hunting operation:

> A: Lynx hunting has had a lot of publicity and media attention. (...) We used to hunt lynx all through the winter season before the quota system was introduced, and back then, teams were never more than about six, eight, or ten men, who hunted the lynx and, well—not many lynx were shot . Not many lynx were shot in Trysil in those years. Granted, the lynx population was perhaps smaller than it is now, possibly because hunting was spread over a longer time, so more lynx were shot maybe during a winter, but there was never a media circus about hunting lynx. Now, there's a lot of attention, and a negative focus on lynx hunting because it lasts just this one day, you know, and a lot of hunters want to get a bit of the action, and they have only one day. When they used to hunt all winter, hardly anyone wanted to join in: it was cold, there was a lot of snow and it was February, it's [devilish] heavy work. But taking part in the hunt on that one day, it's become attractive, that's clear enough.
>
> B: In a way there's this negative slant on [how it is reported], even if it's very exciting and challenging, even if there's a lot of snow and the weather's cold and whatever.
>
> C: So maybe hunting for us hunters is as exciting as it was in those days, but in the media, it's attracted a lot of negative attention because of the number of hunters.
>
> (Hunters group 2, Trysil)

That lynx hunting has a problematic "image" was confirmed in several of our interviews where informants sometimes used terms less than flattering to describe it. In light of the quotation above, we can link this to how the hunt is organized, which, according to the hunters, is necessary because of current hunting regulations. Here is a less positive view of lynx hunting:

> I think [B] and I come at this from slightly different angles; it's got something to do with my general approach to hunting. So, this business about thirty hunters who circle in a lynx they've located and then let the dog loose, or drive it out some other way, and they stand there and shoot this animal—I've no stomach for that type of hunting. I don't even think it is hunting. I hunt moose in an area, for example, where the moose have a pretty fair chance of escaping. (...) The sort of hunting [the lynx hunters do], which in a way is to slaughter [the animal], it's a bit like—for me hunting is a way of harvesting resources and providing food, and the animal should have a chance of getting away. This merciless driven hunting for lynx, which some people are doing, I can't accept it at all, really.
>
> (Trysil municipal council employee)

HUNTING OR CULLING?

Some interest groups (not least within the farming sector) have long held that hunting teams paid by the government should cull or "take out" problem individuals—wolves, bears, wolverines, and lynx. Local hunters could (or preferably should) do the work, but they will have to be paid by the government so that it does not cost local people anything. And the responsibility for large carnivore management is evident: the government—"society at large"—wants large carnivores in Norway, so the government should pay whatever the management costs. Managing large carnivores also involves efficiently eliminating problem individuals.

This line of thought derives from a perception of carnivore management as a technical exercise, the objective of which is to minimize livestock damage. Reducing levels of conflict by measures that result in a wider acceptance of large carnivores is not at all the goal. Advocates of the approach arguably disregard the conflict mitigating potential of ordinary hunting, since ordinary hunting can be attractive and encourage hunters to be more well-disposed toward large carnivores. There is much to indicate that the popular lynx hunting has already had such an effect. But whether this is considered a positive outcome will depend on several factors, including the degree to which a greater carnivore acceptance is seen as desirable. If not, the effect is of dubious value—especially if it reduces the efficiency of targeted removal of alleged problem individuals (which ordinary hunting would have to replace). Our interviews with farmers and hunters reveal somewhat divided opinions on this question. Farmers clearly tended to prefer culls organized and paid for by authorities, if necessary with technology not permitted for hunting under normal conditions:

> A: But I want to return to something else; I don't know if it's the right way of going about it, to send in a culling team, or whatever it is they're using to remove these animals. Why not use the most efficient [technology] we have? We've got helicopters ... Because this isn't hunting. When we're faced with a situation where the harmful animal must be killed, you have to use whatever means and the most appropriate [technology], and that means, well, helicopters or something with an ability to move rapidly and cover a large area in a relatively short time.

> B: I asked [name redacted] to do something about this once, I was so frustrated. I said, "You'll have to start the helicopter and get into the forests and kill this critter." "Impossible," he said. "Why?" I said. "Tagging bears isn't impossible, so it can't be impossible to kill one." "But it's summertime," he said, "so we can't see it!"

> A: Well, there you are, that's what I mean; there's no real will to get rid of problem animals. It's all show, and if there was a real will, obviously they could

find the resources, whether it was professional people with dogs they could fly in from wherever they live, to effectuate the permit. But I wouldn't call it hunting; it's only getting rid of a harmful individual. Hunting doesn't come into it.

B: Putting it down in an efficient, fast way.

[Everyone in unison:] Yes!

Interviewer: But what do you think *Dagbladet* [Oslo tabloid newspaper] would write on its front page, and what sort of light would it put sheep farming in?

A: That's the thing, you see—that's the real problem. [They'll write about how] we are carrying out the gruesome task with helicopters and one thing after the other, but if there was real determination in Norway for a policy to remove harmful animals, politicians or whoever it was who made the decisions would have to show that they have some guts! But no one's got the guts in this country anymore; they say whatever they think people want them to say, and that's what's so hopeless.

(Farmers group 1, Trysil)

Toward the end of this quotation, our informant admits the grazing industry has an "image problem," but the media are to blame, along with (gutless) politicians and "society at large". It goes to show the gravity of the problems facing the sector, say the farmers.

Although many hunters place responsibility for the presence of wolves in Norway at the government's door and believe the government should take responsibility for controlling the wolf population as required, hunters were generally receptive to the idea of hunting the wolf themselves. They argued it would control populations more efficiently than using appointed culling teams. According to hunters, this type of hunting is already popular, despite strict regulations, for example, in the form of quotas and hunting areas. Wolf hunting is currently very limited, but, as we have seen, several hunters believed it could become attractive:

A: If we're going to take out predators, I think it's all right to use hunters. You saw the wolf hunt a few years ago where they used a helicopter, and it doesn't paint a very nice picture of the hunt, nor of hunters in general. It's so wrong somehow, because people who don't know better, they maybe see it as hunting and think that's how hunting is. It gets completely [distorted].

Interviewer: It went much better the next time, to put it like that – and there was much less noise.

B: Not to mention how much cheaper it must have been.

Interviewer: I imagine so, but many people are really annoyed because the government should've paid—local people shouldn't have had to have anything to do with it since it's the government and society at large that want the wolves here, so the government should take responsibility for the hunt. A lot of people in the sheep and livestock industry share that opinion.

B: But that's maybe just to, well, provoke the government, so they can see that it costs something to have [large carnivores] in the forests. I think it's the angle—they want to show that if [the government] want to force this upon us, then it's going to cost them. But when it comes to the actual culling and suchlike, I think local people are probably best.

(Hunters group 2, Trysil)

Interviewer: [Concerning the wolf hunt in (Area X): Opinions were divided about whether it was a good idea to let local hunters do the job, but many believed it was the most effective approach.]

A: Local people are familiar with the area, and—

B: Yes, if we just look at that bear, there was a hunter down the hill here who took a shot at this bear and injured it; then there was another one with dogs who could track it an hour later. But they didn't give him permission; they had to wait more than a day till they came from [more than 400 kilometers away] with [certified] dogs. By then, the scent of the bear was gone.

C: It's no fun going there [with a wounded bear around]. We hunted there, and it was not pleasant. So you pack everything up and go home.

D: And then there's the humane aspect, as well. A bear is wounded, and there are professional people who've been prowling in the forests with moose hounds all their lives. They're just as professional as those others, maybe more professional, too, [yet] weren't allowed—

A: —to go into [the forest]. And then they arrive the day after!

B: When the scent is gone. There was just this pool of blood where he'd winged it, and they found blood many other places as well.

A: Yes, they were finding blood all night.

D: It bled—bled for twelve hours, that bear.

(...)

C: But maybe the police—they don't want people going in after it because of the risk of accidents, that might be the reason. But what do I know? It may have been what stopped them, but twenty-four hours later, it's getting a bit late.

(Hunters group I, Trysil)

Hunters were less interested in removing specific problem animals. To the extent that they touched on the problems affecting the sheep farmers, they believed general population control would reduce the number of attacks because the number of large carnivores would be lower and the animals react to being hunted by keeping their distance. As a hunter from Trysil said: "It's really very logical. If you don't hunt wild animals or pursue them for two hundred years, then they'll lose the shyness they're supposed to have, obviously. Clear as a bell, really."

SUMMING UP

Formal organizational aspects of large carnivore management do not generally seem to interest people who feel affected by carnivore-related issues. For example, most informants had little knowledge of and little interest in issues related to how the management of large carnivores is organized, including decision-making procedures and regional management boards. We often heard people express a desire to have more local power in decision-making (especially in connection with culling permits), but very few knew much about the role of the local authorities or had any practical ideas about how it should be done. We often heard people say, "it's been decided," but without indicating who had made the decision or how the decision-making procedures work. People talked as if decisions were taken far away and high up. The power structures that exist in this field seem very diffuse to many. Indeed, the structure of the wildlife management institutions (as of government bodies in general) is complex and constantly changing. To get an overview of how it works and to follow the changes are both difficult, which is probably why people often have no clear opinions on the structure of the large carnivore management system in Norway. Moreover, few care how the management is organized, as long as it is the practical consequences that affect them. Regardless of people's attitudes toward large carnivores and their management, most people with a clear view on the situation link these consequences to elevated political levels—that is, to the general policy on large carnivores—and not to what are perceived to be organizational details that make no difference in the bigger picture. This seems to be the case regardless of people's attitudes toward large carnivores and their management. Those with opinions often point to decisions taken "up there" and would rather not waste time discussing a complex topic of little importance to them.

People perceive government policy toward large carnivores in a larger social context, which is precisely the subject of this book. Current large carnivore policy can be understood as an element in the ongoing attack on rural economy and lifestyles from "society at large," led by the urban middle class, or conversely as part of flawed environmental policies in which economic interests (in this case farming) always trump concerns for biodiversity and a viable nature. Few see the regional boards or other institutional arrangements as relevant factors in this connection. During the interviews, conversations moved away quickly from these subjects and over to what really concerned people: the general political level and practical local effects. People were likely to show great interest in and detailed knowledge of the practical aspects of large carnivore management that affected them personally. If people hunt large carnivores, they will also have a grasp of the relevant rules and of which agencies are responsible for what. Farmers who have lost sheep are well versed in the workings of compensation schemes and the government agencies responsible for them. In other words, informants care most about the concrete and the immediate—the things that affect them in their everyday lives—and from that angle, the multitude of procedures in government agencies seem more like an impenetrable cacophony of details.

At the same time, the interviews revealed a strong interest in carnivore politics at a general level. While informants were seldom interested in discussing the organization of regional and local management, they willingly delved into questions like why Norway needs wolves and who should manage large carnivores—things that engage people with an interest in the issue. Many see the national political decision to host viable populations of large carnivores in Norway, including wolves, as an attack on rural communities' economic and social foundation. The identity of the "carnivore contact" in the district, and whether the politically appointed regional boards do their job as intended, are in this sense less important questions since they make no difference to national carnivore policies. Many would like local authorities to have more power because, they believe, it would result in larger hunting quotas and faster processing of culling permit applications. This standpoint seems closely related to perceptions of large carnivore populations as considerably larger than official estimates, and only more hunting and faster removal of problem animals can keep populations at reasonable levels. It also seems to rely on an assumption that the current municipal decision-makers agree with this assessment and would have allowed the removal of more animals had they been able to do so.

Whether this removal would have happened is an empirical question nobody can answer today. But it is interesting to observe that the municipalities of Trysil and neighboring Engerdal had a conditional standing permit for the removal of problem bears at the time of our interviews (a trial arrangement that was later abandoned). To the great frustration of sheep farmers, the Trysil municipal council never put the permit into effect. Some informants pointed out that certain council officials had "politically correct" views on the matter of large carnivores, that is, they endorsed the views of conservationists. Accordingly, delegating powers to local authorities would not necessarily ensure a more liberal removal regime. But this did not affect informants' desire for a fully local management system, and evidence was presented to show that the authorities in Engerdal had actually used their "part" of the removal permit more actively. Nevertheless, it points to an interesting dilemma for the advocates of local management: unless it produces tangible results, a local management system is pointless. In other sectors of public administration, where local authorities do have considerable power, there is no ubiquitous confidence in the local councils, to put it mildly. We can only mention policy areas such as senior citizen care, building permits, traffic solutions, public art, and so on. There is probably no reason to expect the local large carnivore management to be more universally popular, especially because many people in every municipality will be sympathetic toward large carnivores and do not want weaker protection.

Our interviews indicate that many residents in areas affected by large carnivores are frustrated over their inability to influence decisions made at the national level. The interviews provide a basis for suggesting that the real distribution of power in large carnivore management creates a sense of powerlessness among people who live in areas with large carnivores, despite attempts to devolve decision-making to the affected areas through regional management bodies. This is partly expressed through the general lack of interest in and knowledge of the specific management arrangements. Formal decision-making at regional and local levels emerges as particularly irrelevant for two reasons. First, important policy guidelines are not laid down here. Second, local- and experience-based knowledge appears to have little real effect on formal decision-making processes. Both points explain some of the widespread discontent with the current system and why it sometimes seems as if everything that originates from the government is doubted and criticized. They also explain why local opponents and supporters of large carnivores seem to share large parts of the criticism that has emerged, which reminds us once again that the entire conflict panorama is influenced by factors that have little to do with large carnivores.

Although large carnivores attract a good deal of attention in political debates and the media, large carnivores do not rank among the top political issues in Norway—or anywhere else. Many are indifferent to the whole issue, even in areas with large carnivores (cf. Skogen 2001). However, many participants in our studies belonged to groups where clear views on large carnivores prevail. They often found themselves in situations where they were obliged to take a stand on the issues raised, and then a different picture emerged. Supporters and opponents of large carnivores—or more precisely, supporters and opponents of populations at current levels—shared many opinions on the government's role. They often concluded that the knowledge underpinning policies and management practices lacks credibility and is therefore untrustworthy.

A key insight when it comes to trust is that an individual who puts trust in someone simultaneously transfers influence to that person. Trust thus becomes an important aspect of power, namely the aspect that concerns legitimate power or authority—a form of power accepted by those subject to it, because they accept the basis on which it is exercised. Authority in this sense involves the ability of those in power to put their will into effect without meeting resistance, and this ability declines when trust is eroded. "Withdrawal of trust by many persons at once—a contradiction of trust—sharply reduces the potential for action of those who had been trusted" (Coleman 1990: 195). James S. Coleman's words point to the possible consequences of mistrust for those in power. Translated into the topic of this book, we may say that government agencies' capacity to act is constrained. Put bluntly, the legitimacy on which the government's exercise of power (read: management of large carnivores) relies crumbles under the weight of the mistrust we find expressed in our material: authority dissipates. The only option remaining for the government is the exercise of unmasked power, meaning power that provokes resistance. The situation entails conflict, which is precisely what characterizes the relationship between many of our informants and the institutions that manage large carnivores in Norway.

The management of large carnivores faces a problem of legitimacy in some segments of the population. Lack of legitimacy is not a trivial problem. If a norm or law lacks legitimacy, people will not feel obliged to comply with it. There are many sectors in society where quite a few people consider prohibitions and injunctions meaningless, and breaking them may be seen as morally justified—in fact, disregarding them may almost be seen as an obligation. To find examples of areas where legislation is considered illegitimate is easy. No direct comparison is intended, but we can mention things such as undeclared work, tax evasion, speeding, and smuggling small amounts of alcohol. Of course, many people support current legislation and may even want tighter regulations. Others, however, and for various reasons, consider them unfair, unreasonable, and perhaps harmful to private enterprise. Perhaps legislation and management related to large carnivores are becoming areas where segments of the population consider the exercise of power illegitimate. This would clearly hamper dialogue, which is problematic enough. A more extreme consequence, however, might be justification for poaching large carnivores. According to biologists, illegal hunting is the most likely cause for 50 percent of wolf mortality in Scandinavia (Liberg et al. 2011). Our studies have not found evidence of widespread illegal hunting, but many clearly understand why it could happen, and over the years we have often heard people say they would never consider reporting such incidents. As we have also seen, this is even truer for actions meant to avert attacks on livestock or dogs. Legally, killing a protected carnivore is a serious offense—and many perceive it as such—but many in the affected communities also perceived prosecution as completely unreasonable.

Since legitimacy challenges are not unique to the management of large carnivores, they cannot be eliminated by efforts restricted to this policy area. Again we return to the social context in which the management of large carnivores and related conflicts unfold. In this sense, the carnivore "field" closely resembles other controversial policy areas, where conflicts to a considerable extent reflect the same social cleavages.

NOTES

1. The seven reproductions that occurred in 2015 (seven litters were born) could be said to disprove this claim. Biologists estimate that the zone could hold twenty packs if this were politically feasible. However, the notion that the zone is too small is part of a view that sees the current management regime as hostile to wolves.
2. High-profile Norwegian wolf biologist.
3. Norway has a "right to roam" legislation, based on historical access rights. This means that access, and even camping and some harvesting, is open to everyone everywhere on uncultivated private land. The land inside the fence is private in a legal sense, but the informant is talking about a ta moral violation of the access right (see also Øian and Skogen 2015).
4. St.meld. nr. 15 (2003–2004) "Rovvilt i norsk natur."

CONCLUDING NOTES

✳ ✳ ✳

A main message in this book posits that conflicts over large carnivores are about a lot more than the animals, which becomes particularly clear when studying the conflicts over wolves. Despite the relative absence of livestock in areas with an established wolf population in Norway, the conflicts have been intense. Hunting with dogs in areas with wolves is difficult, and some people do not like having them close to home. While these are important issues, our focus lies elsewhere. We have shown that the wolf is inscribed into preexisting societal cleavages that reach far beyond management of wildlife. Our work also explains what a study of the wolf conflicts can tell us about general social mechanisms in modern societies, including those connected to class, power relations, and social change.

In chapter 3, we saw how a particular social construction of a rural community, which defines the city and urbanity as its antithesis, takes the edge off internal conflicts and cultural disparities. In the following chapters, we highlighted such disparities, especially related to class. To conclude our sociological journey through the home range of the Norwegian wolf, we aim to bring the two dimensions together—class and cultural cleavages on the one hand and the urban-rural axis on the other. The urban-rural dimension of the large carnivore conflicts is arguably an important component of the social construction of rural communities as being under threat. We have explained the significance of maintaining a boundary against what is outside the community. Here, the "city" and "society at large" are important contrasts—and clear adversaries—in relation to the rural community. For many rural people, the city and urbanity—and not least what is perceived as urban culture (which seeps into the rural communities and makes things even worse)—are important parts of their self-understanding as an image of what they *are not*.

We are not saying wolf conflicts are class conflicts in disguise and completely detached from rural-urban cleavages. If something appears to be real, it tends to have real consequences. If a basic premise of people's thoughts and actions is a notion of a fundamental urban-rural divide, then this perceived antagonism could have, for example, political implications. Furthermore, tangible differences obviously exist between town and country. Problems with wolves materialize in rural areas and many people simply do not relish the idea of being near wolves. People who actually live within the wolf range know what it is like to have wolves as next-door neighbors. Political power, on the other hand, is concentrated in cities. The institutions, organizations and indeed social segments associated with the exercise of power and imposition of regulations that restrict the exploitation of natural resources and other ways of using the land (for example snowmobiling, a hugely controversial issue in Norway) are more likely to be supported by urban people and typified by the middle-class culture of the highly educated, the very people who populate these institutions of power. In a certain sense, modern environmentalism *is* a product of urban culture and is still driven by social segments whose primary allegiances—cultural and economic—are not with the traditional countryside.

The rural perspective on the large carnivore conflicts gains further sustenance from the powerful Norwegian farming sector. The rural policy of Norwegian governments has been built mainly around farming, mostly small-scale, as an attempt to sustain settlement in rural areas and all over the country. This is in stark contrast to the situation in Sweden, where the larger part of the transboundary wolf population resides. There, the smallholders are more or less history, and their numbers in historically feudal Sweden never matched those in Norway anyway. But conflicts over wolves in the thinly populated Swedish countryside are just as serious as in Norway, and according to recent studies, public opinion regarding wolves varies along the same axes in both countries (Krange et al. 2017). Despite these similarities, the large carnivore question has not captured the attention of national political circles in Sweden the way it has done in Norway (although this is now changing, as mentioned in the introduction). Finally, Norway and Sweden have different large carnivore management regimes (for example, Sweden has more than 3,000 bears compared to Norway's 130). In Norway, the hegemonic definition of the large carnivore situation is as a problem essentially for livestock production, and since farming is construed—and politically treated—as a pillar of the Norwegian countryside, large carnivore problems can easily be subsumed as a challenge to rural life in general.

The strength of the farming sector is derived, moreover, from a large bureaucratic apparatus and powerful trade associations. There are well-established procedures for transferring economic support from the government and running various practical schemes. In the case of large carnivores, the sector's strong position is reflected in the priority given to compensation and damage prevention measures, designed primarily to resolve problems related to livestock. This includes the wolf areas, where in fact livestock production is very limited and rough grazing occurs only in a few enclaves. But the strong political position of agriculture and the focus on livestock issues regarding mitigation efforts become drivers leading to unintended consequences that demonstrate the complexity of the large carnivore conflicts. Many practical measures, intended as conflict mitigation, can aggravate both friends and foes of large carnivores, especially those who are not farmers themselves.

The fact that government rangers (SNO personnel) and semi-professional culling teams appointed by municipalities are responsible for the—sometimes extensive—culling of large carnivores, means ordinary hunters have less opportunity to harvest these species. Even so, the culling of "nuisance animals" mainly reduces the overall number of large carnivores. According to ecological research, all large carnivores are potential livestock depredators, so identifying problem individuals is often extremely difficult or impossible (Herfindal et al. 2005; Odden et al. 2002). Regular hunting methods can be used to control population numbers just as effectively, a useful approach since it cultivates a view of large carnivores as ordinary wildlife species. This is precisely what many would like to see, including the Norwegian Association of Hunters and Anglers, although not the people who want large carnivores completely removed—for example some representatives of farming organizations who see this as a dangerous path to acceptance, and thus to defeat

Electric fences are used to prevent depredation, but some of them may hinder the movement of wildlife and impede people's access to large areas. They may also harm the environment when roads are built to run parallel to the fences. Some local people, not least hunters, are provoked by the fencing projects, as we saw in chapter 8. Population monitoring at the level of accuracy required by Norwegian large carnivore policy means many individuals must be immobilized, fitted with GPS devices and moved. Both supporters and opponents of wolves in Norway emphasize the ethical problems and animal welfare concerns involved. These practices also threaten the representation of the wolf as a wild animal (chapter 5). What hunters and other outdoor enthusiasts—whether or not they call themselves environmentalists—want to enjoy in the wild does not include monitored and "domesticated" animals. For some, this amounts to destroying the wild animals they want to experience. For others, the wolf cannot remain in Norway precisely because the type of wilderness habitat wolves need does not exist here anymore, which is why they must be so strictly controlled. In sum, the management system and many of the mitigation efforts seem absurd to both friends and foes of the wolf, and consequently, conflicts are aggravated rather than reduced.

As we have mentioned several times (in chapters 5 and 8, for example), there is no consensus concerning wolves among people who live close to them. Many have accepted that the wolf has returned and even see it in a positive light:

as a natural part of local wildlife, long lost, and as a sign that nature is capable of recovery. Indeed, some hunters and farmers also share this view. But farming and natural resource industries only employ a minority of rural people today. Social groups without strong connections to traditional land use are growing in rural areas. In many cases, they are newcomers, often belonging to the middle class, who are carriers of academic knowledge and associated with cultural and political dominance. Many of the people living in sparsely populated areas have made a deliberate but sometimes demanding choice to settle there, though it is rarely the easiest option. Some have dug in where they grew up; some have moved to the country to enjoy close encounters with the wilderness. People who appreciate nature this way will often find value in every living species, including large carnivores. They may feel ambivalent about having predators in their backyard, not to mention in their neighbor's sheep herd, but they believe these animals have a right to exist as an intrinsic part of the natural environment in which they live.

The patterns we describe here are not causal connections. Many people living in rural areas are highly educated yet negative toward wolves. These include not only the "modern" landowners and sheep farmers but also people without any economic interests threatened by wolves. And, as we said in chapter 3, all segments of the local population may be drawn toward the idea of a strong, cohesive community. This latter point brings us to an aspect of the wolf conflict that has been present throughout this book. To a large degree, these conflicts have a class dimension—not primarily the old contradiction between labor and capital (even if that also plays a part) but rather between labor and abstraction. This cleavage can overlap the urban-rural divide, but only partly, and probably to a decreasing extent. Cleavages like these are part of the wider pattern of cultural realignment and conflict typical of the times we live in. We would particularly highlight the growing influence of science and emergence of formalized knowledge systems and institutions. They have been essential to the rapid expansion of the middle class over the past 150 years and its current influence. The modern middle class is large and powerful, a social force of increasing strength across a widening front.

In connection with the development of the Norwegian welfare state, postwar governments, mostly from the Labor Party, built a massive bureaucracy, where scientific knowledge was intended to support planning and social engineering. This applied to all areas of public policy, and it was well intentioned. The ever-expanding bureaucracy of policy areas such as health, social services, culture, and eventually the environment, was aimed at serving the public, and was populated by highly qualified members of the middle class. The close relationship between the middle class and the modern state is evident not least in the research sector. Many research institutions started their life as investigative departments in the national directorates. A connection—and convergence of interests—exists here between those who manage hegemonic knowledge and the modern state apparatus.

For a long time, through the postwar period, the labor movement managed to speak on behalf of "ordinary people," despite the creation—by the ruling Labor Party—of strong government institutions advocating scientific knowledge in many areas and the powerful position, within the party itself, of strong individuals who had a background in, for example, medicine and economics. A basis for this was the struggle against another elite, the economic upper class. As the labor movement itself became associated with both political power and academic knowledge, it grew increasingly difficult to sustain this image. And while economic dominance has a greater impact on how people may live their lives, cultural dominance is easier to see. Today, people's irritation with know-it-alls and pesky experts who want to tell them what to do in almost every area of human existence tend to overshadow economic power and injustice. But the cultural hegemony of the middle class, which is closely connected to political influence, is facing resistance. It is seen as arrogant and intrusive, and "ordinary people" will not stand for it. There has been a long process during which a growing segment of the public has revisited its opinion of parties that once defended working people's interests. They think of them now as elite-dominated organizations, where the reins of power are in the hands of small groups originating in the highly educated middle class. This is not a uniquely Norwegian phenomenon, and not restricted to large carnivore territory, as can be clearly seen from cultural and political shifts currently occurring around the globe.

Distrust of science is far from new and has always been a fundamental characteristic of working-class culture. It has been the cornerstone of many ethnographic studies of the industrial working class. In a Norwegian context, Sverre Lysgaard's 1950s classic *Arbeiderkollektivet* (The workers' collective) comes to mind. Paul Willis's contributions from England convey the same message: the cultivation of practical skills and technical prowess has always been a source of self-respect, particularly for working-class men. The ridicule of academic knowledge that goes with it has also reflected workers' actual experience of having impractical, desktop plans foisted on them by engineers and managers (Lysgaard [1961] 1985; Willis 1977, 1979). Newer research has shown that these staple cultural elements are still very much alive, even among young people (Borgen and Skogen 2013; McDowell 2003; Nayak 2006).

We have observed that a deep mistrust of academic knowledge is very much alive among working-class Norwegians. In our field of study, this is expressed in relation to wildlife management, land use, and conservation. Many still consider academic knowledge and research little more than airy speculation in the service of special interests and should be no match for well-founded, practical, everyday knowledge. The cultural dominance of the middle class is extra annoying to many people because the type of knowledge it represents is so conspicuously linked to the exercise of power.[1] As mentioned, this is not to say people who see themselves in a subordinate position do not put up a resistance. However, this resistance is not primarily directed against the old upper class, but rather against the middle class and their cultural dominance. This entails resistance against social groups who see it as their mission to protect the natural environment, the air we breathe, public health, architecture, art, and children—and who enjoy a discursive hegemony in these areas, although with variable real-world impact (as any architect would testify). They are also firmly positioned within formal structures of power. Opposition to this type of hegemony converges in an alliance that supersedes the class divisions described by so many sociologists and political activists, from Karl Marx on. Put bluntly, we can say capitalist and worker have come together to fight patronizing and sanctimonious middle-class experts whose political correctness stands in the way of economic progress, profits, jobs—and the dignity of working-class people.

The construction of the tightly knit rural community—understood as having common interests across social groups—is the rural version of this alliance, which has a material base as well as a cultural superstructure. Defending the rural culture and traditional land use (or rather certain constructions of rural culture and land use) have become cornerstones of a cultural resistance with a distinct class dimension. Like many other forms of cultural resistance, it shifts the front lines from one type of class-related conflict to another, which is where wolves come in and play an important role. They do not create these conflicts, nor do they drive the construction of social cohesion. New forms of class conflicts as well as social constructions of community are found in places with no wolves and no other contentious wildlife. But the wolf can help us understand important conflict dimensions and processes of change in contemporary society. And it is impossible to explain human-wolf conflicts without taking the wider social context into account.

NOTE

1. Obviously, scientific "truths" and their application by the authorities and powerful economic actors are criticized from many quarters. Groups within the middle class itself voice much of this criticism, not least in the fight against the perceived trivialization of environmental risks, as discussed in chapter 6. While this is important, we concentrate here on a main topic of the book, which deals with another type of resistance against the hegemony of scientific knowledge and its social basis.

BIBLIOGRAPHY

Aagedal, O., and Å Brottveit. 1999. *Jakta på elgjaktkulturen*. Oslo: Abstrakt forlag.

Almås, R. 1989. "Characteristics and conflicts in Norwegian agriculture." *Agriculture and Human Values*, 6(1), 127–136.

Abric, J.-C. 1984. "A Theoretical and Experimental Approach to the Study of Social Representations in a Situation of Interaction." In R. Farr and S. Moscovici (eds.), *Social Representations*. Cambridge: Cambridge University Press.

Abric, J.-C. 1993. "Central System, Peripheral System: Their Functions and Roles in the Dynamics of Social Representations." *Papers on Social Representations*, 1(1): 75–78.

Andersen, R., J.D.C. Linnell, and H. Hustad. 2003. *Rovvilt og samfunn i Norge: En veileder til sameksistens i det 21. århundre*. Trondheim: Norsk institutt for naturforskning.

Arnot, M. 2004. "Male Working-Class Identities and Social Justice: A Reconsideration of Paul Willis's Learning to Labor in Light of Contemporary Research." In N. Dolby and G. Dimitriadis (eds.), *Learning to Labor in New Times*. New York: RoutledgeFalmer.

Artdatabanken. 2015. *Rödlistade arter i Sverige 2015*. Uppsala: ArtDatabanken SLU.

Bauer, M. W., and G. Gaskell. 2008. "Social Representations Theory: A Progressive Research Programme for Social Psychology." *Journal for the Theory of Social Behaviour* 38(4): 335–353.

Bauman, Z. 2000. *Liquid Modernity*. Cambridge: Polity Press.

Beck, U. 1995. *Ecological Politics in an Age of Risk*. Cambridge: Polity Press.

Beck, U. 2000. *Risk Society: Towards a New Modernity*. London: Sage.

Bell, M.M. 1994. *Childerley: Nature and Morality in a Country Village*. Chicago: University of Chicago Press.

Berger, P.L., and T. Luckmann. 1967. *The Social Construction of Reality*. London: Penguin.

Bertaux, D., and P. Thompson. 1997. *Pathways to Social Class: A Qualitative Approach to Social Mobility*. Oxford: Clarendon Press.

Bjerke, T., O. Reitan, and S.R. Kellert. 1998. "Attitudes toward Wolves in Southeastern Norway." *Society and Natural Resources* 11(2): 169–178.

Bjerke, T., K. Skogen, and B. Kaltenborn. 2003. *Nordmenns holdninger til store rovpattedyr: Resultater fra en spørreskjemaundersøkelse*. Lillehammer: Norsk institutt for naturforskning.

Blekesaune, A., and K. Rønningen. 2010. "Bears and Fears: Cultural Capital, Geography and Attitudes towards Large Carnivores in Norway." *Norsk Geografisk Tidsskrift: Norwegian Journal of Geography* 64(4): 185–198.

Boglioli, M. 2009. *A Matter of Life and Death: Hunting in Contemporary Vermont.* Amherst: University of Massachusetts Press.

Borgen, O.Å., and K. Skogen. 2013. "Gutta på jakt: Jakt som arena for reproduksjon av arbeiderklassekultur." *Tidsskrift for ungdomsforskning* 13(1): 3–30.

Bourdieu, P., and J.B. Thompson. 1991. *Language and Symbolic Power.* Cambridge: Polity Press in association with Basil Blackwell.

Bourgois, P. 2003. *In Search of Respect.* Cambridge: Cambridge University Press.

Braverman, H. 1974. *Labor and Monopoly Capital: The Degradation of Work in the Twentieth Century.* New York: Monthly Review Press.

Brottveit, Å., and O. Aagedal. 1999. *Jakta på elgjaktkulturen.* Oslo: Abstrakt forlag.

Campion-Vincent, V. 1976. "Les histoires exemplaires." *Contrepoint* 22–23: 217–232.

Campion-Vincent, V. 1990. "Histoires de lâchers de vipères: Une légende francaise contemporaine." *Ethnologie française* 20: 143–155.

Campion-Vincent, V. 2000. "Les réactions au retour du loup en France: Une analyse tenant de prendre 'les rumeurs' au sérieux." *Anthropozoologica* 32: 33–59.

Campion-Vincent, V. 2004. "The Return of the Wolf in France." *Journal of Indian Folkloristics* 5 (1–2): 133–162.

Campion-Vincent, V. 2005a. "From Evil Others to Evil Elites." In G.A. Fine, V. Campion-Vincent, and C. Heath (eds.), *Rumor Mills.* New Brunswick, NJ: AldineTransaction.

Campion-Vincent, V. 2005b. "The Restoration of Wolves in France: Story, Conflicts and Uses of Rumor." In A. Herda-Rapp and T.L. Goedeke (eds.), *Mad about Wildlife: Looking at Social Conflict over Wildlife.* Leiden: Brill, pp. 99–122.

Chambre d'agriculture des Alpes Maritimes. 1996. *Loups et élevage: Une cohabitation impossible.* Nice: Chambre d'agriculture des Alpes Maritimes.

Cohen, A.P. 1985. *The Symbolic Construction of Community.* Chichester: Ellis Horwood.

Coleman, J.S. 1990. *Foundations of Social Theory.* Cambridge, MA: Belknap Press.

Coombe, R.J. 1997. "The Demonic Place of the 'Not There': Trademark Rumors in the Postindustrial Imaginary." In A. Gupta and J. Ferguson (eds.), *Culture, Power, Place.* Durham, NC: Duke University Press, pp. 249–274.

Cotgrove, S., and A. Duff. 1980. "Environmentalism, Middle-class Radicalism and Politics." *Sociological Review* 28(2), 333–351.

Crow, G., and G. Allen. 1994. *Community Life: An Introduction to Local Social Relations.* New York: Harvester-Wheatsheaf.

Dickens, P. 1996. *Reconstructing Nature: Alienation, Emancipation, and the Division of Labour.* New York: Routledge.

Dolby, N., and G. Dimitriadis. 2004. *Learning to Labor In New Times.* New York: RoutledgeFalmer.

Douglas, M. 1984. *Implicit Meanings.* London: Routledge.

Douglas, M. 1992. *Risk and Blame: Essays in Cultural Theory.* London: Routledge.

Douglas, M. 2002. *Purity and Danger.* London, New York: Routledge.

Douglas, M., and A. Wildavsky. 1982. *Risk and Culture: An Essay on the Selection of Technical and Environmental Dangers.* Berkeley: University of California Press.

Dressel, S., C. Sandström, and G. Ericsson. 2015. "A Meta-analysis of Studies on Attitudes toward Bears and Wolves across Europe 1976–2012." *Conservation Biology* 29(2), 565–574.

Dunk, T. 1991. *It's a Working Man's Town: Male Working-Class Culture in Northwestern Ontario.* Montreal: McGill-Queen's University Press.

Dunk, T. 1994. "Talking about Trees: Environment and Society in Forest Workers' Culture." *Canadian Review of Sociology and Anthropology* 31(1): 14–34.

Dunk, T. 2002. "Hunting and the Politics of Identity in Ontario." *Capitalism, Nature, Socialism* 13(1): 36–66.

Eder, K. 1993. *The New Politics of Class: Social Movements and Cultural Dynamics in Advanced Societies.* London: Sage Publications.

Ericsson, G., and T.A. Heberlein. 2003. "Attitudes of Hunters, Locals and the General Public in Sweden Now that the Wolves Are Back." *Biological Conservation* 111(2): 149–159.

Evans, G. 2006. *Educational Failure and Working Class White Children in Britain.* Houndsmill: Palgrave MacMillan.

Farrugia, D. 2013. "Towards a Spatialised Youth Sociology: The Rural and the Urban in Times of Change." *Journal of Youth Studies* 17(3): 293–307.

Fegan, B. 1986. "Tenants' Non-Violent Resistance to Landowner Claims in a Central Luzon Village." *Journal of Peasant Studies* 13: 97–106.

Félonneau, M.L. 2003. "Les représentations sociales dans le champ de l'environnement". In G.G. Moser and K. Weiss (eds.), *Espaces de vie.* Paris: Armand Colin, pp. 145–176.

Figari, H., and K. Skogen. 2011. "Social Representations of the Wolf." *Acta Sociologica* 54(4), 317–332.

Fine, G.A., and P.A. Turner. 2001. *Whispers on the Color Line.* Berkeley: University of California Press.

Frank, T. 2004. *What's the Matter with Kansas? How Conservatives Won the Heart of America.* New York: Metropolitan Books.

Garde, L. 1998. *Loup et pastoralisme: La prédation et la protection des troupeaux dans le contexte de la présence du loup en Région Provence-Alpes-Côte-d'Azur.* Manosque: Centre d'Etudes et de Réalisations Pastorales Alpes Méditerranée.

Gaskell, G., and M.W. Bauer. 1998. "The Representations of Biotechnology: Policy, Media and Public Perceptions." In J. Durant, M. Bauer, and G. Gaskell (eds.), *Biotechnology in the Public Sphere: A European Source Book.* London: Science Museum Press, pp. 3–12.

Gervasi, V., E.B. Nilsen, H. Sand, M. Panzacchi, G.R. Rauset, H.C. Pedersen, and J.D.C. Linnell. 2012. "Predicting the Potential Demographic Impact of Predators on their Prey: A Comparative Analysis of Two Carnivore-ungulate Systems in Scandinavia." *Journal of Animal Ecology* 81(2): 443–454.

Giddens, A. 1991. *Modernity and Self-identity: Self and Society in the Late Modern Age.* Cambridge: Polity Press.

Gjems, S.R., and E. Reimers. 1999. *ABC for jegerprøven.* Oslo: Landbruksforlaget.

Gullestad, M. 1984. *Kitchen-table Society.* Oslo: Universitetsforlaget.

Gupta, A., and J. Ferguson. 1997. *Culture, Power, Place: Explorations in Critical Anthropology.* Durham, NC: Duke University Press.

Hall, S. 1996. "Gramsci's Relevance for the Study of Race and Ethnicity." In D. Morley and D. K. Chen (eds.), *Stuart Hall.* London: Routledge.

Hebdige, D. 1979. *Subculture: The Meaning of Style.* London: Methuen.

Herfindal, I., J.D.C. Linnell, P.F. Moa, J. Odden, L.B. Austmo, and R. Andersen. 2005. "Does Recreational Hunting of Lynx Reduce Depredation Losses of Domestic Sheep?" *Journal of Wildlife Management* 69(3): 1034–1042.

Herzlich, C. 1973. *Health and Illness: A Social Psychological Analysis.* London: European Association of Experimental Social Psychology by Academic Press.

Jodelet, D. 1984. "The Representation of the Body and Its Transformations." In R. Farr and S. Moscovici (eds.), *Social Representations.* New York: Cambridge University Press.

Jovchelovitch, S. 2008. "The Rehabilitation of Common Sense: Social Representations, Science and Cognitive Polyphasia." *Journal for the Theory of Social Behaviour* 38(4), 431–448.

Kapferer, J.-N. 1990. *Rumeurs: Le plus vieux média du monde.* Paris: Seuil.

Krange, O., C. Sandström, G. Ericsson, and T. Tangeland. 2017. "Attitudes towards Wolves in Scandinavia: A Comparison between Norway and Sweden." *Society and Natural Resources.*

Krange, O., and K. Skogen. 2007a. "Kodebok for den intellektuelle middelklassen." *Nytt norsk tidsskrift* (3): 227–242.

Krange, O., and K. Skogen. 2007b. "Reflexive Tradition: Young Working-Class Hunters between Wolves and Modernity." *Young: Nordic Journal of Youth Research* 15(3): 215–233.

Krange, O., and K. Skogen. 2011. "When the Lads Go Hunting: The 'Hammertown Mechanism' and the Conflict over Wolves in Norway." *Ethnography* 12(4): 466–489.

Kriesi, H. 1989. "New Social Movements and the New Class in the Netherlands." *American Journal of Sociology* 94(5): 1078–1116.

Lareau, A. 2003. *Unequal Childhoods: Class, Race and Family Life.* Berkeley: University of California Press.

Liberg, O., H. Andrén, H.-C. Pedersen, H. Sand, D. Sejberg, P. Wabakken, and S. Bensch. 2005. "Severe Inbreeding Depression in a Wild Wolf *Canis lupus* Population." *Biology Letters* 1(1): 17–20.

Liberg, O., G. Chapron, P. Wabakken, H.C. Pedersen, N.T. Hobbs, and H. Sand. 2011. "Shoot, Shovel and Shut Up: Cryptic Poaching Slows Restoration of a Large Carnivore in Europe." *Proceedings of the Royal Society B: Biological Sciences* 279.

Liepins, R. 2000. "Exploring Rurality through 'Community': Discourses, Practices and Spaces Shaping Australian and New Zealand Rural 'Communities.'" *Journal of Rural Studies,* 16(3): 325–341.

Linnell, J., and T. Bjerke. 2002. "Frykten for ulven: En tverrfaglig utredning." *NINA Report.* Trondheim: NINA.

Lysgaard, S. [1961] 1985. *Arbeiderkollektivet.* Oslo: Universitetsforlaget.

Martin, B. 1998. "Knowledge, Identity and the Middle Class: From Collective to Individualised Class Formation?" *The Sociological Review* 45: 653–686.

Mauz, I. 2005. *Gens, cornes et crocs: Relations hommes-animaux et conceptions du monde, en Vanoise, au moment de l'arrivée des loups.* Paris: Cemagref, Cirad, Ifremer, Inra.

McDowell, L. 2003. *Redundant Masculinities? Employment Change and White Working Class Youth.* Malden, MA: Blackwell.

Merton, R. K. 1995. "The Thomas Theorem and the Matthew Effect." *Social Forces* 74(2): 379–422.

Milgram, S. 1984. "Cities as Social Representations." In R. Farr and S. Moscovici (eds.), *Social Representations.* New York: Cambridge University Press, pp. 289–309.

Moscovici, S. 1963. "Attitudes and Opinions." *Annual Review of Psychology* 14: 231–260.

Moscovici, S. 1969. "Preface to Herzlich." In C. Herzlic (ed.), *Santé et maladie: Analyse d'une représentation sociale.* Paris: Edition de l'Ecole des Hautes Etudes en Sciences Sociales.

Moscovici, S. 1976. *La psychanalyse, son image, son publique.* Paris: Presses Universitaires de France.

Moscovici, S. 1993. "Des représentations collectives aux représentations sociales." In D. Jodelet (ed.), *Les représentations sociales.* Paris: Presses Universitaires de France, pp. 62–86.

Moscovici, S. 2001. "The phenomenon of social representations." In G. Duveen (ed.), *Serge Moscovici: Social representations.* New York: New York University Press, pp. 18–77.

Naughton-Treves, L., R. Grossberg, and A. Treves. 2003. Paying for Tolerance: Rural Citizens' Attitudes toward Wolf Depredation and Compensation. *Conservation Biology* 17(6): 1500–1511.

Nayak, A. 2006. "Displaced Masculinities: Chavs, Youth and Class in the Post-industrial City." *Sociology: The Journal of the British Sociological Association* 40(5): 813–831.

Nelson, P.B. 2001. "Rural Restructuring in the American West: Land Use, Family and Class Discourses." *Journal of Rural Studies* 17(4): 395–407.

Odden, J., J.D.C. Linnell, P. Fossland Moa, I. Herfindal, T. Kvam, and R. Andersen. 2002. "Lynx Depredation on Domestic Sheep in Norway." *The Journal of Wildlife Management* 66(1): 95–105.

OECD. 2012. *Agricultural Policy Monitoring and Evaluation 2012.* Paris: OECD Publishing.

Øian, H., and K. Skogen. 2015. "Property and Possession: Hunting Tourism and the Morality of Landownership in Rural Norway." *Society and Natural Resources* 29(1): 104–118.

Ortner, S. 1995. "Resistance and the Problem of Ethnographic Refusal." *Comparative Studies in Society and History* 37: 173–193.

Räikkönen, J., A. Bignert, P. Mortensen, and B. Fernholm. 2006. "Congenital Defects in a Highly Inbred Wild Wolf Population (*Canis lupus*)." *Mammalian Biology* 71(2): 65–73.

Samper, D. 2002. "Cannibalizing Kids: Rumor and Resistance in Latin America." *Journal of Folklore Research* 39(1): 1–32.

Scott, J.C. 1990. *Domination and the Arts of Resistance: Hidden Transcripts.* New Haven, CT: Yale University Press.

Skogen, K. 1999. "Another Look at Culture and Nature: How Culture Patterns Influence Environmental Orientation among Norwegian Youth." *Acta Sociologica* 42(3): 223–239.

Skogen, K. 2001. "Who's Afraid of the Big, Bad Wolf? Young People's Responses to the Conflicts over Large Carnivores in Eastern Norway." *Rural Sociology* 66(2): 203–226.

Skogen, K. 2003. "Adapting Adaptive Management to a Cultural Understanding of land Use Conflicts." *Society and Natural Resources* 16(5): 435–450.

Skogen, K., and O. Krange. 2003. "A Wolf at the Gate: The Anti-carnivore Alliance and the Symbolic Construction of Community." *Sociologia Ruralis* 43(3): 309–325.

Skogen, K., and O. Krange. 2010. "Middelklassemakt? Nei takk!" In J. Ljunggren and K. Dahlgren (eds.), *Klassebilder.* Oslo: Universitetsforlaget, pp. 157–168.

Skogen, K., I. Mauz, and O. Krange. 2008. "Cry Wolf! Narratives of Wolf Recovery in France and Norway." *Rural Sociology* 73(1): 105–133.

Skogen, K., and C. Thrane. 2008. "Wolves in Context: Using Survey Data to Situate Attitudes within a Wider Cultural Framework." *Society and Natural Resources* 21(1): 17–33.

Skuland, S.E., and K. Skogen. 2014. "Rovdyr i menneskenes landskap." *Tidsskriftet Utmark* 14 (1–2). URL: http://utmark.nina.no/Portals/utmark/utmark_old/utgivelser/pub/2014-1%262%26S/ordin/Skuland_Skogen_UTMARK_1%262.html

Snerte, K. 2001. *Ulvehistorier.* Oslo: Det norske samlaget.

Statistics Norway (SSB). 2004. "Fra skuddpremier til fredning og irregulær avgang." www.ssb.no/vis/emner/historisk_statistikk/artikler/rovdyr/art-2004-05-05-01.html.

Strandbu, Å., and O. Krange. 2003. "Youth and the Environmental Movement: Symbolic Inclusions and Exclusions." *Sociological Review* 51(2): 177–198.

Strandbu, Å., and K. Skogen. 2000. "Environmentalism among Norwegian Youth: Different Paths to Attitude and Action." *Journal of Youth Studies* 3: 189–209.

Tangeland, T., K. Skogen, and O. Krange. 2010. *Om rovdyr på landet og i byen: Den urban-rurale dimensjonen i de norske rovviltkonfliktene.* Oslo: Norsk institutt for naturforskning.

Toverud, L. 2001. *Utsetting av ulv i Norge og Sverige 1976–2001.* Bjørkelangen: Lars Toverud.

Vila, C., A.K. Sundqvist, O. Flagstad, J. Seddon, S. Bjornerfeldt, I. Kojola, H. Ellegren. 2003. "Rescue of a Severely Bottlenecked Wolf (*Canis lupus*) Population by a Single Immigrant." *Proceedings of the Royal Society of London Series B: Biological Sciences* 270 (1510): 91–97.

Wabakken, P., L. Svensson, E. Maartmann, M. Åkesson and Ø Flagstad. 2016. *Bestandsovervåking av ulv vinteren 2015–2016: Bestandsstatus for store rovdyr i Skandinavia 1.* Evenstad and Grimsö: Rovdata and Viltskadecenter, SLU.

Wagner, W., and N. Hayes. 2005. *Everyday Discourse and Common Sense: The Theory Of Social Representations.* New York: Palgrave Macmillian.

Weis, L. 1990. *Working Class without Work.* New York: Routledge.

Willis, P. 1977. *Learning to Labour: How Working Class Kids Get Working Class Jobs.* Aldershot: Gower.

Willis, P. 1979. "Shop Floor Culture, Masculinity and the Wage Form." In J. Clarke, C. Critcher, and R. Johnson, (eds.), *Working Class Culture.* London: Hutchinson.

Wilson, M.A. 1997. "The Wolf in Yellowstone: Science, Symbol, or Politics? Deconstructing the Conflict Between Environmentalism and Wise Use." *Society and Natural Resources* 10(5): 453–468.

Wright, E.O. 1997. *Class Counts: Comparative Studies in Class Analysis.* Cambridge: Cambridge University Press.

Wynne, B. 1996. "May the Sheep Safely Graze? A Reflexive View of the Expert-Lay Knowledge Divide." In S. Lash, B. Szersynski, and B. Wynne (eds.), *Risk, Environment and Modernity: Towards a New Ecology.* London: Sage.

INDEX

❄ ❄ ❄

www.ingramcontent.com/pod-product-compliance
Lightning Source LLC
Chambersburg PA
CBHW070925030426
42336CB00014BA/2533